TIME
LIFE BOOKS ®

Other Publications:

THE ENCHANTED WORLD
THE KODAK LIBRARY OF CREATIVE PHOTOGRAPHY
GREAT MEALS IN MINUTES
THE CIVIL WAR
PLANET EARTH
COLLECTOR'S LIBRARY OF THE CIVIL WAR
LIBRARY OF HEALTH
CLASSICS OF THE OLD WEST
THE EPIC OF FLIGHT
THE GOOD COOK
THE SEAFARERS
WORLD WAR II
THE OLD WEST
LIFE LIBRARY OF PHOTOGRAPHY (revised)
LIFE SCIENCE LIBRARY (revised)

For information on and a full description of any of the
Time-Life Books series listed above, please write:

Reader Information
Time-Life Books
541 North Fairbanks Court
Chicago, Illinois 60611

This volume is part of a series offering homeowners
detailed instructions on repairs, construction
and improvements they can undertake themselves.

HOME REPAIR
AND IMPROVEMENT

ADVANCED WIRING

BY THE EDITORS OF
TIME-LIFE BOOKS

TIME-LIFE BOOKS
ALEXANDRIA, VIRGINIA

HOME REPAIR AND IMPROVEMENT

Editorial Staff for Advanced Wiring

Editor William Frankel
Assistant Editor Mark M. Steele
Designer Anne Masters
Picture Editor Adrian G. Allen
Associate Designer Kenneth E. Hancock
Text Editors William H. Forbis, Jim Hicks, Brian McGinn,
Ellen Phillips
Staff Writers Thierry Bright-Sagnier, Stephen Brown, Alan Epstein,
Steven J. Forbis, Lydia Preston, Brooke C. Stoddard,
David Thiemann
Art Associates George Bell, Michelle Clay, Mary Louise Mooney,
Lorraine Rivard, Richard Whiting
Editorial Assistants Megan Barnett, Eleanor G. Kask

Editorial Operations
Design Ellen Robling (assistant director)
Copy Room Diane Ullius
Production Anne B. Landry (director), Celia Beattie
Quality Control James J. Cox (director), Sally Collins
Library Louise D. Forstall

Correspondents: Elisabeth Kraemer-Singh (Bonn);
Margot Hapgood, Dorothy Bacon (London); Miriam
Hsia, Susan Jonas, Lucy T. Voulgaris (New York); Maria
Vincenza Aloisi, Josephine du Brusle (Paris); Ann
Natanson (Rome). Valuable assistance was also provided
by: Judy Aspinall, Lesley Coleman, Karin B. Pearce
(London); Carolyn T. Chubet, Christina Lieberman (New
York); Mimi Murphy (Rome).

THE CONSULTANTS: Kent P. Stiner, an electrical engineer, is a former member of the Correlating Committee of the National Electrical Code and has served on code drafting panels since 1946. He has been chief electrical inspector of the city of Detroit, manager of product planning and engineering for an electrical-equipment manufacturer and a teacher of electrical design at Michigan State University. He has published articles in professional journals on electrical inspection and on residential and commercial wiring.

Lawrence E. Kennedy is an electrical contractor and master electrician. He teaches electrical wiring at Thomas A. Edison High School, Alexandria, Virginia, and in adult education and veterans' courses. During World War II he served as an electrical shop chief for B-17 bomber maintenance.

Harris Mitchell, special consultant for Canada, has worked in the field of home repair and improvement for more than two decades. He is Homes editor of Today magazine and author of a syndicated newspaper column, "You Wanted to Know," as well as a number of books on home improvement.

Roswell W. Ard, a civil engineer, is a consulting structural engineer and a professional home inspector.

Contents

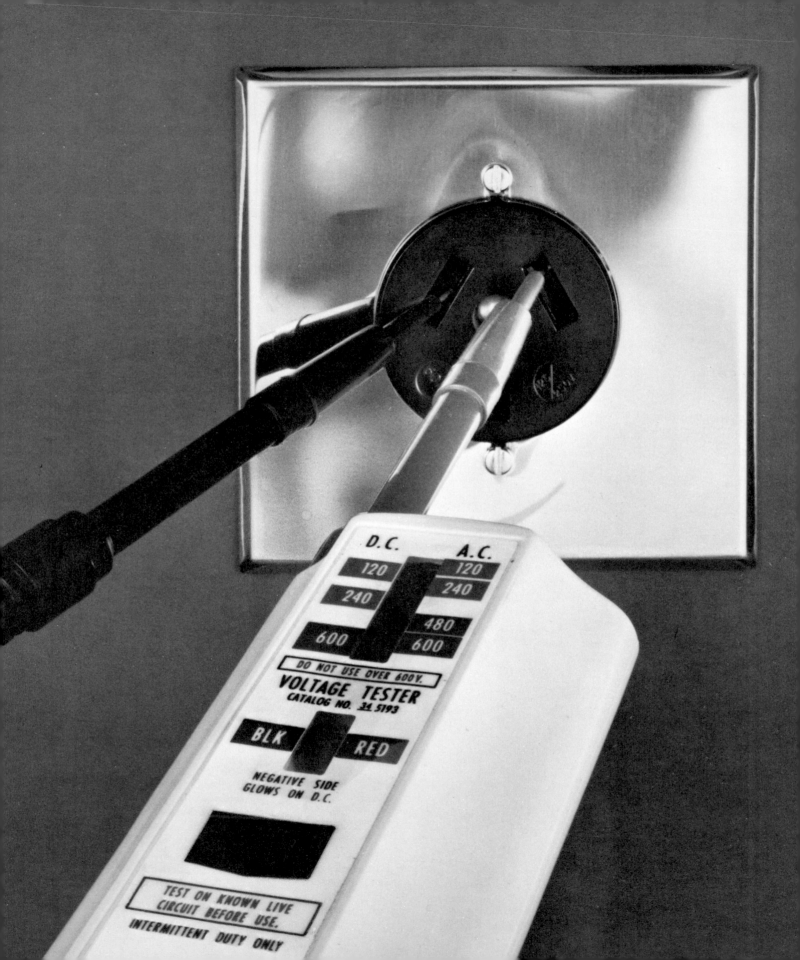

1

Wiring Modern Power and Control

Testing the work. A bar-meter voltage tester, used here to check a 240-volt receptacle, provides an essential measure of safety by ensuring that the outlet is correctly wired before any appliance is plugged in. Other testers, shown on pages 8 and 37, locate short-circuits before power is restored and diagnose wiring hazards that might otherwise go undetected.

A few decades ago, the only home electrical loads of any consequence were lights, a toaster and perhaps a fan or two. A 120-volt supply and 30 amperes of current provided enough power for almost any home; the only control devices were a few wall switches. After World War II, 240-volt service was added to 120, and higher amperages powered an array of appliances—clothes washers and driers, kitchen appliances and many more. Toggle wall switches replaced the push-button type and circuit breakers superseded fuses.

In most homes, that is where matters stand today. Home builders generally have kept up with a growing need for power, but they have not given the homeowner the benefit of modern advances in low-voltage control and communication. It is now possible to control lights and receptacles from any number of remote-control switches, from a bedside master panel—or, automatically, from a timer or a photoelectric cell. In a high-fidelity system you can install remote speakers throughout the house; you can improve your television system with an antenna rotator; and you can put in low-voltage intercoms and even a private telephone system.

The owners of older homes often have a special problem: they simply do not have enough power. New appliances cannot be installed without overloading a circuit, and some circuits are already overloaded, with fuses that blow or breakers that trip several times a month. These problems can be solved by running new circuits from the service panel, using basic wiring techniques to connect fixtures and receptacles and applying the techniques shown in the following chapter to make the final hookup at the panel.

Working Safely with Electricity

The jobs described in Chapters 1 and 2 of this book—connecting new circuits at the service panel and upgrading the service of an entire house—deal with potentially dangerous current. But the jobs are no more dangerous than such basic wiring tasks as replacing a wall switch—if you take reasonable precautions. To protect yourself from injury you must follow these rules:
☐ Always shut off power upstream before you begin work. If you are working on a branch circuit, turn off the circuit breaker or pull the fuse—and label the service panel so no one will restore power. If you are working at the service panel, have an electrician pull the meter (page 50) or have the utility company disconnect the power.
☐ Before you begin work, check the body of a fixture and every possible combination of wires with a voltage tester to make sure power is off. If you find voltage or notice wires with damaged insulation, call in an electrician.
☐ Never work near live power. If you drill through a wall, fish wires between floors, or work near a service drop, turn off power in the vicinity.
☐ If your job is covered by an electrical permit, do not turn power on until an inspector approves your work.

Tools to Speed Electrical Work

Although the wiring techniques described in this book go beyond the routine installation of an outlet or replacement of a light fixture, they do not necessarily require tools other than those in the homeowner's basic kit. However, the more specialized tools pictured opposite will speed and simplify advanced operations—particularly those involved in work on service panels and subpanels. These tools include cutters and rippers for large cables and conduit, drills for masonry, and special cable-pullers as well as a number of testers that are more accurate and more versatile than the neon glow lamp ordinarily used.

☐ TESTERS. A bar-meter voltage tester measures voltages ranging from 120 to 600 volts. Its spring-loaded, retractable prongs can be inserted into the slots of a receptacle or fixed in the open position for voltage tests in a panel or meter box. When extended from the tester body, the prongs can take a voltage reading on terminal lugs that are widely separated. The tester is used to make sure power has been shut off at a panel or a circuit and to check the wiring of 240-volt circuits.

A low-voltage tester, which takes readings between 5 and 50 volts, checks stereo and low-voltage switching circuits. The receptacle analyzer, when plugged into the slots of a receptacle, offers a quick way to identify wiring faults. Two lights on the analyzer body go on if the receptacle and its outlet box are correctly wired; other lights diagnose up to five common wiring errors.

A continuity tester, which uses a battery to send a small amount of current through a new circuit, is used to find wiring problems before a circuit is energized and to locate short circuits.

☐ STRIPPERS. Cable rippers, metal clips 4 to 5 inches long with sharp teeth inside their bodies, fit over flexible cable up to ⅝ inch in diameter to slice neatly and quickly through the sheathing. The handle of the ripper contains a wire gauge.

☐ HEX WRENCHES. Terminals for heavy cable are often fitted with hexagonal-head screws in several different sizes. Hex-wrench sets keep as many as nine different wrenches at hand and the large handles provide leverage for loosening or tightening the screws.

☐ PIPE CUTTER. To fit a number of lengths of heavy-wall conduit, you may wish to rent a pipe cutter. Though a hacksaw will do the job, the cutter will save time and make neater cuts. Two carbide wheels in the head of the tool cut a groove as the tool is rotated around the conduit; they are tightened with each revolution until they cut clear through.

☐ WIRE BASKET. The most convenient way to secure wire and cable to the end of a fish tape is with a wire basket. A cylinder woven of thin steel strands, it squeezes tight around a wire in its mouth when the tape is pulled.

☐ FISH TAPE. A fish tape with a grip built into the body is superior to conventional tapes for pulling service cable or thick bunches of wire around corners and through sharp conduit bends. The body of the spool turns independently of the handle for extra pulling power and the handle can be used to rewind the tape.

☐ ELECTRICIAN'S KNIFE. This folding knife has a safety lock to keep it from closing in use. The hooked portion of the blade is particularly useful for stripping insulation in the tight confines of a panel or meter box; the straight part of the blade is used to strip insulation from heavy cable before it is connected.

☐ FASTENERS. Because a hammer drill operates without electricity, it can be used when current has been turned off for work on circuits. Carbide-tipped bits, which drill neat holes in masonry, are available in sizes to match the lead shields designed to anchor large panel boxes. For mounting lighter objects—cable clamps or small electrical boxes—to masonry, a drive-pin set is useful. The nail-like steel pins, ½ to 1½ inches long, fit into the barrel of the tool and are driven into the wall when the piston is hit with a heavy hammer.

☐ CUTTERS. Diagonal-cutting pliers have short cutting jaws and long handles for maximum leverage. Available in two sizes, they make short work of slicing through cables up to No. 2 in size. For larger cables, use the large model to snip through a few strands at a time, or use a hacksaw fitted with a blade with 32 teeth per inch. Use the small size to cut wires in cramped quarters, such as the interior of a panel box.

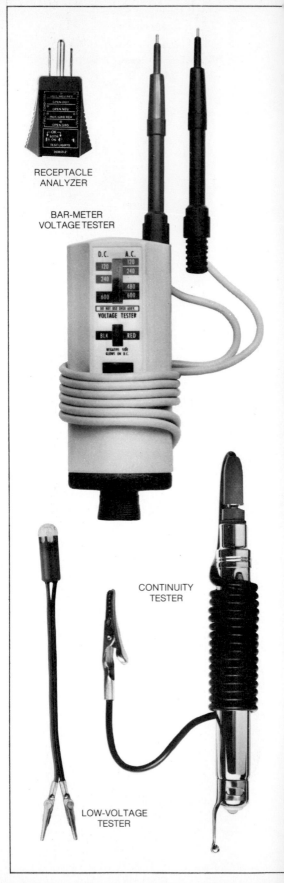

RECEPTACLE ANALYZER

BAR-METER VOLTAGE TESTER

CONTINUITY TESTER

LOW-VOLTAGE TESTER

CABLE RIPPERS

HEX WRENCH SETS

PIPE CUTTER

N° 3-0 HEAVY DUTY
1" TO 3"

WIRE BASKET

DRIVE-PIN SET

HAMMER DRILL

BITS

ELECTRICIAN'S
KNIFE

FISH TAPE

DIAGONAL-CUTTING PLIERS

Getting the Current Wherever You Need It

The kind of cable you need for a wiring job depends on three factors: the purpose of the circuit (do you intend to connect the house to utility lines or bring power to a lamp?), the number of amperes of current to be carried, and the location of the cable (inside walls, underground or buried in concrete?). A multitudinous special class of cables handles low voltages for stereo sets and intercoms and is discussed in Chapter 3. However, for power supplies—120 and 240 volts—and for lights and appliances, discussed in this and the following chapter, only a few types of cable are important:

□ Service entrance concentric (SEC) cable is used between overhead power company lines and a house's electric meter. It also connects the meter to either a service panel's fuses or circuit breakers or to a main disconnect—a master switch, mounted in a box separate from the panel, that can turn off all the current in the house. The cable is called concentric because the bare strands of wire that serve as its neutral conductor are wrapped completely around an inner core consisting of two black insulated hot wires. All of these wires may be either aluminum or copper—though aluminum wires are dangerous in branch circuits, they are as safe as copper in a SEC cable, because the SEC connectors are specially designed to accept them.

□ Service entrance round (SER) cable runs from a main disconnect to a service panel; it is rarely used in a house that does not have a separate main disconnect. SER cable is as large as SEC cable, but has a different make-up: bare ground wires are twisted together into a single thick bundle, and the neutral wire is a single line insulated in white or gray.

□ Underground feeder (UF) cable is sheathed in tough, moistureproof plastic and can be buried in the earth. It comes in an especially wide range of sizes. The smallest is used for individual branch circuits, such as those that power yard lights; larger sizes run between a main service panel and a subpanel in a garage or outbuilding, and the largest of all, called an underground service entrance (USE) cable, connects an electric meter to underground power lines.

□ Type NM cable is the workhorse of modern interior wiring. (The initials stand for nonmetallic, but electricians often refer to NM cable as Romex, the name of the best-known brand.) In relatively small 12- and 14-gauge sizes, this plastic-sheathed cable serves the 15- and 20-ampere branch circuits that supply light fixtures and receptacles for small appliances. In sizes 10-gauge and larger, NM cable serves large appliances and subpanels; these large NM cables, which carry 30 to 70 amperes of current, contain either two or three conductors. The more versatile three-conductor cable, with one white conductor and the others either black or black and red, is most often used as feeder cable for subpanels and for 120/240-volt range circuits: the two hot wires are used together to power the 240-volt heating elements, while the neutral wire and one hot wire supply 120 volts for a range-top clock or light. A large two-conductor cable is used for appliances such as heaters, which require only 240-volt current. In addition, there is nonmetallic cable (NMC), similar in appearance and use to NM cable, but heavily insulated for damp locations.

□ Armored cable, which contains the same sizes of conductors as UF and NM cable, is sheathed with a spiral steel jacket. It is not as flexible as plastic-sheathed cable, but must be used when wiring is buried in mortar or secured to a surface that is especially vulnerable to damage. For damp locations, use a special armored cable called ACL, which has lead-covered conductors.

Once you have chosen the kind of cable you need for a specific wiring job, you can go on to determine the exact size to buy. The maximum current for every size of cable is strictly limited by the electrical code; violating the limit overheats wires and can cause fire. But calculating the amount of current permitted by the code is not a simple matter of adding the amperages of the loads that a cable must serve. In service and feeder cable, for example, such simple addition would call for a larger, more expensive conductor than you need—a home may contain lights and equipment totaling 500 amperes, but only a few are likely to be in use at any time. Therefore, the code prescribes several formulas that discount much of the total load (see page 124).

The code also recognizes that all branch circuits are not alike. It defines them by location and use and prescribes different rules for each. Three kinds are common in houses: the general lighting circuit, the small appliance circuit and the individual appliance circuit. A general lighting circuit (opposite), which powers light bulbs and low-amperage appliances such as record players and clocks, can be wired to more receptacles than a small appliance circuit, which serves such rooms as a kitchen or dining room, where toasters, coffee makers and other high-amperage devices are used. And an individual 120- or 240-volt appliance circuit serves a single outlet, one each for range, water heater, dishwasher or disposal.

Detailed rules for calculating loads and determining wire sizes are given on pages 124-125. You cannot shortcut these rules when sizing service and feeder cables, but several rules of thumb will make the planning of branch circuits easier:

□ General lighting and small appliance circuits can carry 15 or 20 amperes in the United States. In Canada they are limited to 15 amperes unless a single appliance in the circuit requires more than 15 amperes. A general lighting circuit can serve one lighting fixture or single receptacle for each 1½ amperes of its capacity. A small appliance circuit should serve no more than six duplex receptacles; locate at least one receptacle of such a circuit in the dining room to serve toasters and the like. Use 12-gauge cable for 20-ampere circuits; 14-gauge for 15-ampere circuits.

□ An individual appliance circuit should carry enough current to handle the rated load of the appliance, printed on its name plate or given in the owner's manual. For motor-driven appliances, such as an air conditioner, multiply the load by 1¼ to allow for the surge of current when the motor starts. The chart on page 124 indicates the wire sizes for a variety of individual appliance circuits.

When you plan a branch circuit of any type, do not hesitate to add some extra capacity. It takes no longer and is only slightly more expensive to run a 20-ampere circuit rather than a 15-ampere one, and by going beyond the code minimums you can allow for future additions.

An array of cables. Each of these cables has special properties that suit it for its function in a home electrical system. The service entrance cables that bring power to circuit breakers or fuse boxes, both SEC and SER, have thick, tough outer sheathings to resist moisture and fire. The sheathing of UF cable, meant for use underground, is also fungus and corrosion resistant, and fills the space inside the cable so that the individual wires run in separate channels. Non-metallic cable, which is used for most of the cable runs inside the house, combines light-weight plastic sheathing and relatively thin wires to make an inexpensive cable that can easily be snaked inside walls. Armored cable is wrapped with a steel jacket for protection against physical damage in special applications.

While the colors and characteristics of the outer sheathing vary widely from cable to cable, the insulation covering the conductors inside the sheathing follows a single code: hot wires are red or black, neutral wires are white or gray and ground wires are green or bare; the only exceptions to the rules occur in SEC cable, which uses a bare wire for neutral rather than for ground. In some switch circuits, white wire may be hot—but any white wire used in this way must be recoded black with paint or tape wherever it enters a box.

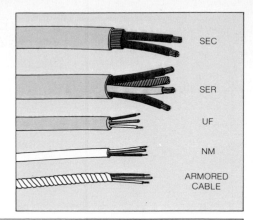

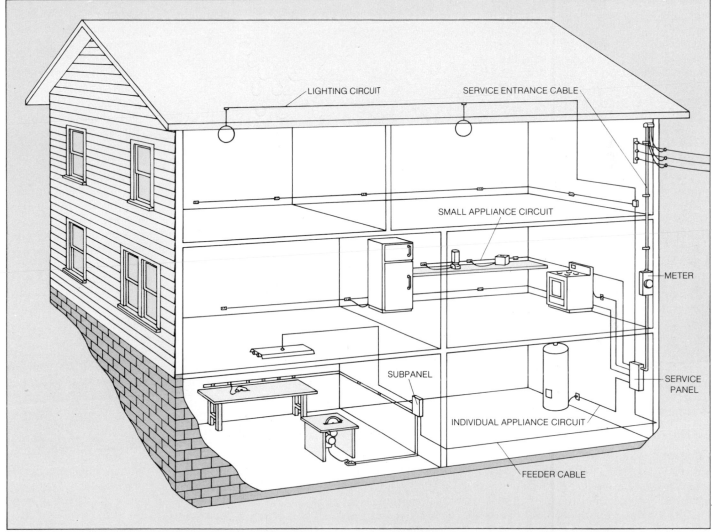

The right cable for every job. The drawing above includes every type of cable likely to be found in a home; the wiring of the circuits has been simplified for clarity. The largest line is the service cable that carries power from the overhead power lines, through the meter, to the service panel in the basement. All the other cables in the house branch out from there. The feeder cable running to the subpanel in the basement workshop is big enough to handle all the loads wired to the subpanel. Two individual appliance circuits serve high-amperage equipment: heavy two-conductor cable supplies 240 volts to the water heater; a three-conductor cable supplies a 120/240-volt circuit to the electric range, which needs 240 volts for its heating elements and 120 volts for its light and clock. Two small appliance circuits, wired with 12-gauge cable to carry 20 amperes, power groups of receptacles; one circuit serves the refrigerator and extends to the dining room, where tabletop appliances might be used; the other provides the kitchen work space with receptacles. A general lighting circuit, the most common type of all, is wired with 14-gauge cable to provide 15 amperes for the upstairs light fixtures and several receptacles.

Professional Tricks for Hiding Cable

A simple circuit extension, made with short runs of cable, can usually be routed through an interior wall, which is hollow. Adding a completely new circuit, on the other hand, or undertaking any extensive rewiring in a house, may mean taking part of a cable run through an exterior wall, which is likely to be packed with insulation, sealed by vapor barriers and blocked by bracing and fire stops—short pieces of wood set horizontally between studs to retard the spread of flame. And some exterior walls are solid masonry.

The easiest way to get past the obstacles of an exterior wall is to avoid them, taking wires up through an unused chimney or heating duct or alongside the plumbing vent "stack," which rises from floor to floor through holes that are generally larger than the stack piping. In a wall of concrete blocks, wire can be fished through the hollow cores of the blocks; if a masonry wall is covered with wallboard nailed to furring strips, you may find a handy space for a horizontal run of cable beneath the bottom strip.

In addition to these tricks, direct attack works on a frame exterior wall. The clogged spaces between its studs are not completely impassable—wire can be run between the vapor barrier and the back of the wallboard. When you hit a fire stop or brace, you can cut a hole through the wallboard and chisel a notch in the obstruction, then cover the wire with a sheet-metal plate and patch the hole. And if long runs or the rustling of the vapor barrier make it difficult to tell when two fish tapes meet, you can wire them to a small electric bell that will ring when the tapes touch.

Once a fish tape extends along to the full length of a cable run, hook cables to it in the conventional way or use a basket (*page 14*) to pull many wires at once. While you pull the tape back, have a helper push the cables or wires at the far end—the extra force will help move them past obstructions. Always pull enough cable to provide at least 6 inches of slack in every junction box, plus plenty of extra for trimming and stripping.

Dodging Obstacles in Exterior Walls

Belling a fish tape. After feeding fish tapes into the wall, hook a terminal of a 6-volt dry cell to one terminal of an inexpensive doorbell. Run an 18-gauge insulated wire from the other battery terminal to one of the fish tapes and fasten it to the tape with an alligator clip. In the same way, run a wire from the other bell terminal to the second fish tape. When the fish tapes touch, the bell will sound.

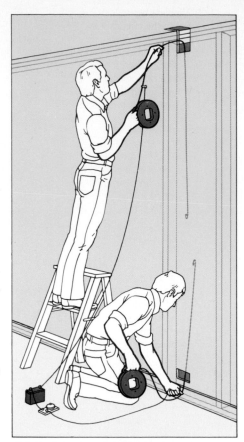

Notching an obstruction. If a fire stop or brace blocks a cable run, cut a hole over the obstruction about halfway between two studs and chisel a notch in the wood large enough to hold the cable without pinching it. Run the cable through the notch; then, seal the notch with a $1/16$-inch metal plate nailed to the fire stop or brace and patch the wall. Most exterior walls contain insulation, omitted here for clarity. You probably will have to push or cut insulation away in order to notch a fire stop or brace.

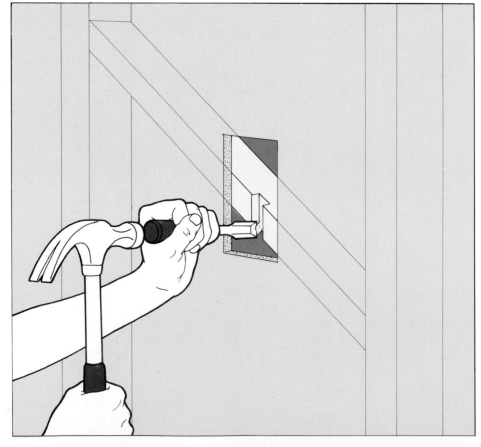

Fishing in concrete block. Punch a hole into a concrete block at the top of a wall with a hammer and cold chisel and push a fish tape down through the hollow cores of the blocks below. If the tape meets an obstruction such as a blob of mortar between two courses, flip the tape in a half circle, using its natural curl to turn the head toward the opposite side of the block. Keep twisting the fish tape until you can hook it to another, inserted from the bottom.

In some houses, one or two courses at the top of a wall are made of solid block. If this is the case, you must use armored cable. Begin fishing through a hole punched into the hollow cores below the solid blocks. Run armored cable in a notch chiseled in the solid blocks, and cover both the cable and the notch with premixed patching mortar (*page 17, top*).

Caution: if you find any signs of moisture on the surface of the wall or inside the blocks, the Code requires cable approved for damp locations. Use NMC or UF nonmetallic sheathed cable (*page 11*); where armored cable is needed use type ACL, which has lead-covered conductors.

Using an I beam. To make a long cable run between opposite masonry walls, you can often take advantage of a boxed I beam; remove a section of wallboard and push a fish tape through. Use the curl of the fish tape to keep the hook against the steel and away from the strips that support the wallboard.

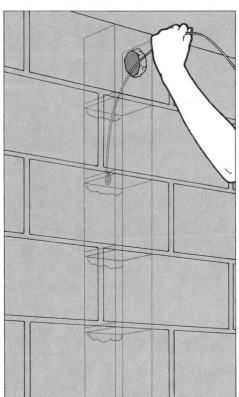

Running below a furring strip. Remove the baseboard and make a small hole in wallboard fastened to masonry by furring strips to see whether the bottom furring strip runs above the level of the floor—carpenters often raise this strip to provide a good nailing surface for baseboards. Snake a fish tape through the space below the strip for a long horizontal run. If the furring strip is not raised above the floor, chisel a groove for the wire in the wallboard. Although the code requires that cable that is run less than 1¼ inches from the back of the wallboard be protected with a metal plate 1/16 inch thick, this requirement is often disregarded. In any case you should renail the baseboard carefully to avoid hitting the cable or plate.

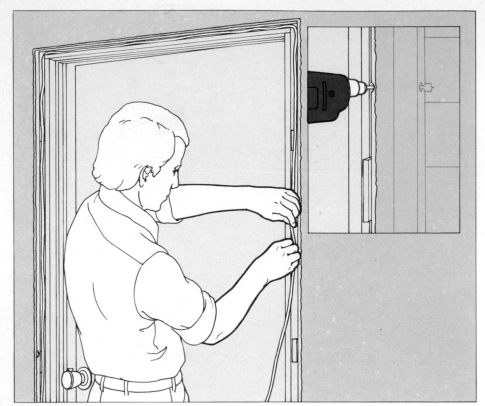

Running around a door. Remove the door-casing covering the gap between the jambs and the rough frame, use a sharp chisel to split away part of the shims that hold the jambs in place and lay in the cable. Caution: when chiseling the shims, be very careful you do not change the relative positions of the paired shims and thus twist the doorframe. If the cable cannot be pushed at least 1¼ inches from the back of the casing, it should, according to the National Code, be covered with a metal plate. When you replace the trim, angle the nails slightly to avoid hitting the cable.

To run a cable outdoors, as for a door-side light, use a long spade bit to drill a hole through the rough frame, between shims, at a 45° angle away from the door (inset).

Fish-tape Connections for Heavy Hauls

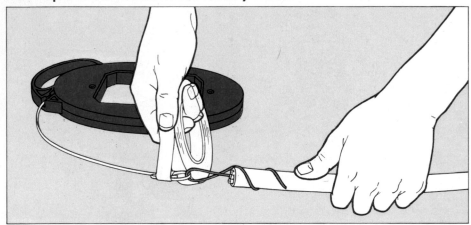

Fishing a large NM cable. To avoid a bulky fish-tape connection that snags at obstructions, cut back the sheathing of a cable and cut off the ends of the hot and the neutral wires, leaving 6 inches of the bare ground wire to make a loop over the fish tape. Twist the end of the loop around the sheathing of the cable, then wind electrical tape over the entire connection, beginning at the fish tape and tapering the turns of tape over the sheathing of the cable.

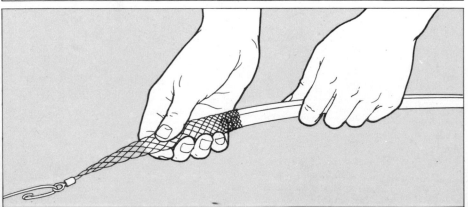

Using a wire basket. To pull several small cables at one time, compress a wire basket (page 9) by pushing the mouth back toward its eye hook and insert the cables. The basket will tighten around the cables as it is pulled.

The Right Way to Strip Cable

NM Cable. Although a knife or pliers can be used to strip NM cable, a cable stripper is better. Close a stripper around the cable so that the tooth pierces the sheathing, then pull hard to slide the stripper to the end of the cable. Peel the sheathing back and cut it off.

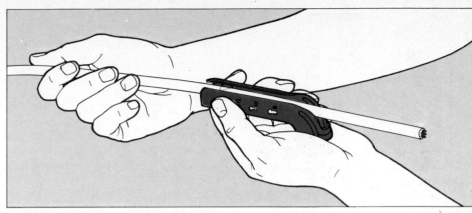

UF cable. Snip 1 inch of cable sheathing open with diagonal cutters and bend the end of the bare ground wire out of the sheathing; then grasp the sheathing with one pair of pliers and the ground wire with another, and pull the ground wire upward to rip the sheathing. Pull the sheathing from each conductor in the same way and cut off the sheathing.

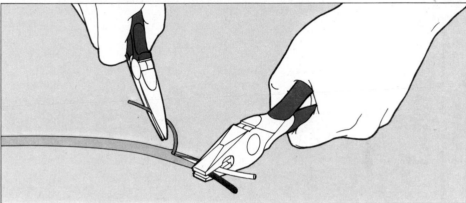

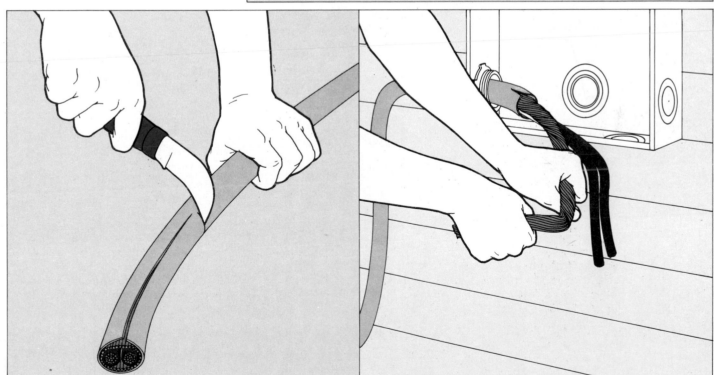

SEC cable. Set the cable on a flat surface and use an electrician's knife to make two cuts—one along the sheathing and the second around the cable at the base of the first *(left)*. Pull the sheathing back and cut it off. After feeding the cable into a panel box, prepare the neutral wires for their connection by bunching them to one side of the cable, then bending 6 inches of these wires at a right angle; use the bent portion as a handle to twist the strands together *(right)*. Cut off the bend with diagonal-cutting pliers, and trim the twisted portion as needed.

Putting Outlets in Tight Places

The basic components of a branch circuit—outlet boxes in which electrical connections are made, and devices such as switches and receptacles that are mounted in the boxes—have been standardized for decades. Nevertheless, innovations are constantly introduced to meet special installation problems. Some of the most important of these innovations involve changes in the size or shape of boxes and devices, in the way boxes are mounted and in the materials of which boxes are made.

The standard outlet box, ranging from 2½ to 3½ inches deep, serves well for most construction and remodeling; the installer simply nails the box to a stud, hangs it between joists or secures it behind walls with tabs or expanding clamps. But a homeowner who wants to locate switches or receptacles in unlikely spots can turn to special boxes—uncommonly shallow ones to fit the thin space between wallboard and masonry; or, when wall surface is at a premium, uncommonly deep ones with room for multiple switches and receptacles.

Special techniques are also needed on some occasions. A standard box fits neatly and easily into a hole in wallboard or plaster; a special one may fit into a notch cut in a stud or a hollow chiseled in brick or cinder block. Because these openings take extra time to make, most builders prefer the standard installation methods, but a homeowner, doing the job himself, can take the time to avoid an inconveniently placed fixture.

Uncommon switches and receptacles are also available to the exacting homeowner. Called interchangeables, they are so small that three can be mounted where only one or two normal devices would fit. Though interchangeables have the same depth as regular devices—codes specify the depth a box must have for a given number of connections—and are wired in the same way, they must be fitted with a special windowed plate called a plaster ring before they can accept a standard receptacle cover. But their small size permits combinations for added convenience.

Before 1970, nearly all outlet boxes were made of metal; today, less expensive plastic boxes are increasingly common. The two materials are interchangeable, but their installation differs. Plastic boxes do not require cable clamps, because there are no shards of metal around the knockout holes to chafe the cables. And because plastic will not conduct electricity, the boxes cannot be grounded; instead, the devices in them are grounded with wire-cap connections and jumper wires *(page 19)*. Caution: to ground a switch correctly in a plastic box you must use a model that comes with a separate grounding connection, identifiable by a green screw terminal. Do not ground a switch with a connection to its strap or mounting screws.

In every other respect, follow the standard procedures for installing a new circuit. Run the cable for the circuit before mounting the boxes, then connect and ground the devices, check the circuit *(pages 36-37)* and connect it to the service panel or subpanel *(pages 24-31)*.

Adapting Boxes to Fit

Notching a stud for a shallow box. Expose the stud with a horizontal hole the same size as the box, and drill two rows of small holes ¾ inch deep into the top and bottom of the exposed wood. Angle the top holes slightly upward and the bottom holes slightly downward. Chisel away the wood between the holes and, after drawing the cables into the box, screw the box to the stud. The box is mounted horizontally so that the cables can clear the stud.

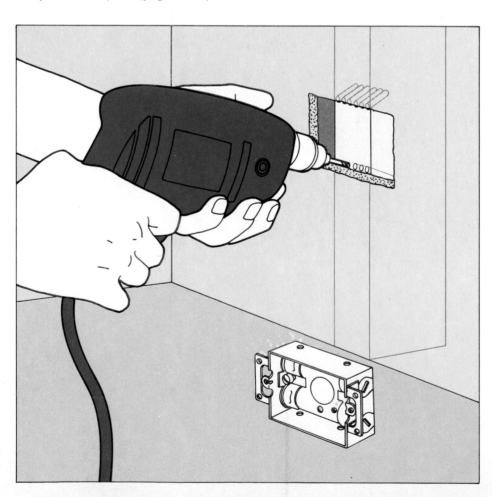

Mounting a box in masonry. Trace the box on the wall and, wearing gloves and safety goggles, chip away the brick inside the line with a cold chisel and a heavy hammer. Wedge the box in place with brick chips (*below, left*). To wire the box use armored cable concealed by mortar. Chisel a V-shaped groove in the brick, secure the cable with a pin and clamp (*page 52, Step 4*), then cover the cable with mortar tinted to match the wall (*right*).

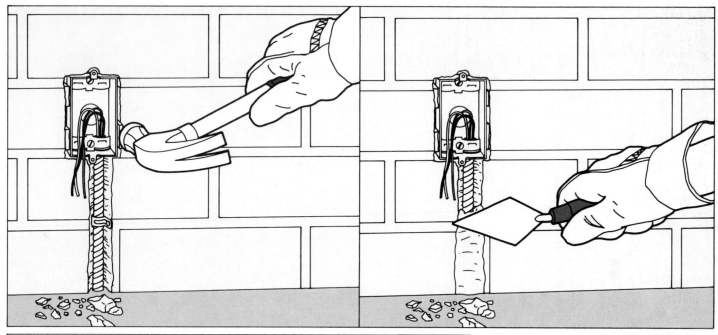

Flush mounting with a plaster ring. For a box that will contain interchangeable devices (*page 18*), cut a hole as wide as the box and 1 inch taller through the wallboard next to a stud, clamp the cables in the box and angle the box, fitted with a plaster ring, into the hole. Bring the ring flush to the surface of the wall and screw the box to the stud. After wiring the devices, fill the gap between the plaster ring and the wall with steel wool and patching plaster.

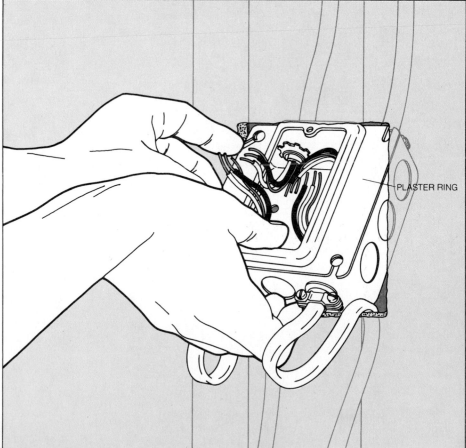

PLASTER RING

Spacesaving Switches and Receptacles

Fastening a device to a strap. Fit the tongue of a strap into the matching slot in a device. With the tip of a screwdriver bend the T slot opposite the tongue to grip the device. To remove the device, twist the T slot in the opposite direction.

Two receptacles and a switch. To control one receptacle with a switch while the other remains supplied with power from a separate circuit, use a wire cap to join the incoming and outgoing black leads of the first circuit (*at right in this example*) and run a black jumper from the cap to one terminal of the switch. Connect the other terminal of the switch to the brass terminal of a receptacle; join the incoming and outgoing neutral leads with a wire cap and connect them to the silver terminal of the receptacle with a white jumper. Wire the other receptacle into the second circuit as you would wire any normal receptacle. Finally, use a large wire cap to join the ground wires of both circuits and run green insulated or bare copper jumpers from the cap to the green ground terminals of the receptacles and to a grounding screw in the box.

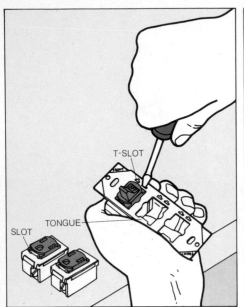

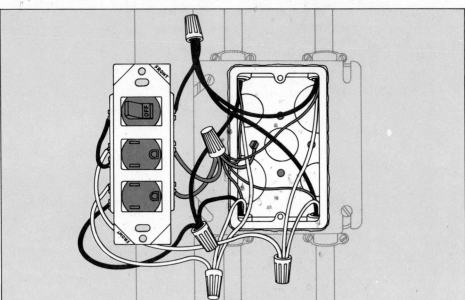

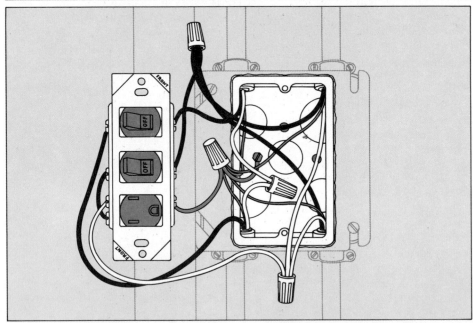

Two switches and a receptacle. In this example, the top switch controls a fixture (*not shown*) in one circuit, while the bottom switch controls a receptacle in the other.

Wire the top switch into the fixture circuit (in this case, the left-hand circuit). Join the incoming and outgoing black leads of the other circuit and run a black jumper from the wire cap to a terminal on the middle switch. Run a second black jumper from the other switch terminal to the brass terminal of the receptacle, then use a wire cap to join the incoming and outgoing neutral leads of the circuit with a jumper wire that connects to the silver terminal of the receptacle. Finally, join all the bare copper wires and link them with a jumper to a grounding screw on the box and the ground terminal of the receptacle.

Grounding Plastic Boxes

In a wall box. Use a wire cap to join the incoming and outgoing ground leads, and run a jumper from the cap to the green grounding terminal on the device—in this example, a switch. The other connections are made in the usual way by attaching the incoming and outgoing black leads to the brass terminals of the switch and joining the incoming and outgoing white leads.

In a ceiling box. Use a wire cap to join the incoming and outgoing ground leads and run two jumpers from the cap—one to the grounding screw on the light fixture, the other to a grounding screw on the mounting bracket of the box. (Some new plastic boxes have grounding straps; if there is one, connect the second jumper to it instead of to the bracket.) Join the incoming and outgoing black leads with one cap and the white neutral leads with another, and run jumpers from the caps to the fixture.

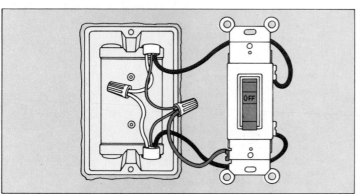

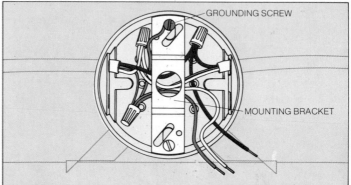

GROUNDING SCREW

MOUNTING BRACKET

The Controversy over Grounding

It was not until the early 1950s that electricians began adding grounds to electrical circuits in any systematic way. Occasionally, conscientious wiremen would wrap a scrap of wire between an outlet box and a water pipe, and in 1955 this practice was made mandatory by the National Electrical Code. Later, the code decreed that any outlet within 8 feet of a cold-water pipe had to be connected to the pipe with a ground wire—but the rule did not spell out how to make the connection. Many electricians tucked the wire under a cable clamp or any other handy protrusion in the box, and taped or twisted it around the pipe.

In the early 1960s both the U.S. and Canadian codes tightened their grounding requirements. Three-hole receptacles with separate grounding terminals were mandated for kitchens, bathrooms and laundries—they are now required in all rooms—and the correct grounding connections were spelled out for the first time. No longer could any nearby pipe be used; the grounds had to terminate in the service panel. And no longer could ground wires merely be twisted together and pushed into a box—they had to be pressure-connected with ei-

ther a wire cap or else a bonding screw.

The choice between a cap and a screw is not as simple as it may seem, for while the U.S. and Canada have standardized their connection methods, the two countries stand in sharp disagreement on this question.

Canada frowns upon the use of wire caps and jumpers; its code suggests that the ground wire be looped around a bonding screw, then connected to the grounding terminal on the device *(upper right)*. The U.S. code requires caps and jumpers. Most electricians believe that both methods work but doubt that the conflict will ever be resolved. The Canadian code writers contend that wire caps can come loose and the pigtailed wires inside them can come apart; their U.S. counterparts argue that a wire grounded to a screw can break.

Whatever the connection method, experts in both countries predict that grounding requirements will continue to grow more stringent. Grounded switches, now used mainly in plastic boxes *(above)*, will probably be required universally, and the use of the ground-fault interrupter (GFI) *(page 44)* will almost certainly be expanded.

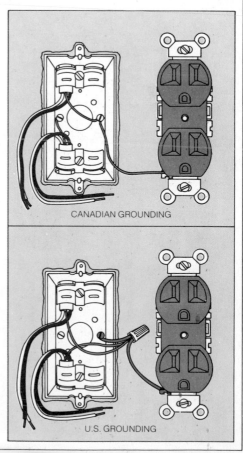

CANADIAN GROUNDING

U.S. GROUNDING

Surface Wiring: The Easiest Way to Add a Circuit

There is no easier way to run a new circuit in your home than by using raceway—interlocking metal channels that run along the outsides of walls and floors, shielding individual hot, neutral and ground wires and housing receptacles and switches. Raceway is best suited for garages, workshops or other dry locations where concealing wires is impractical or where you may want to change or add to a circuit at a later date. The type shown here, derived from electrical hardware originally developed for industry, combines especially high wire-carrying capacity with surprising compactness—the receptacles and switches are located inside the raceway channel, not in full-sized, obtrusive boxes.

Installing this type of raceway system begins with attaching base pieces to the floor or wall. The pieces can be cut with a fine-tooth hacksaw (40 teeth per inch) and come with regularly spaced knockouts for flathead screws. Wires are laid in the base and protective U-shaped wire clips are set over the wires at 1-foot intervals. The switches and receptacles are wired and set in position, and the covers are snapped into place over the base. One type of wall raceway, which comes with receptacles prewired at regular intervals, can be partially disguised by recessing it as a baseboard. In another type, shown on this page and opposite, you can wire in the devices of your choice at any location you please. For both types, special base and cover pieces route raceway around corners and connect it to the electrical system.

You can activate a new raceway circuit from a service panel with a fitting called an offset connector or from an outlet box with a fitting called an adapter plate. Be sure that the power is off when you work at either of these sources. Be sure, too, that the base pieces in the system are securely fastened to one another. This check has electrical as well as mechanical importance: the base pieces are designed to form a continuous ground from every part of the raceway system back to the service panel or outlet box.

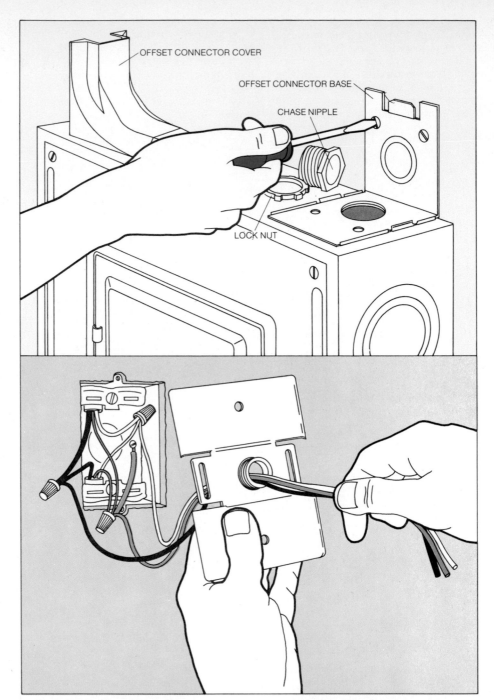

OFFSET CONNECTOR COVER

OFFSET CONNECTOR BASE

CHASE NIPPLE

LOCK NUT

1 **Connectors at a power source.** To connect raceway directly to a service panel (*top*), turn off power to the panel, then remove matching knockouts of the panel box and an offset connector base, and screw the connector base to the wall. Fasten the connector base to the panel box with a type of terminal adapter called a chase nipple and a lock nut. The circuit wires will run through this nipple, but should not be connected at the panel until all other connections have been made.

To tap a receptacle box (*bottom*), turn off power to the circuit, then disconnect the recepta-cle. Using wire caps, join the wires in the box, white to white, black to black and ground to ground; add new lengths of matching wires, each 8 to 10 inches long, at the caps. Thread the three new wires through the raceway adapter plate, push the other connections into the box and screw on the plate. Now thread the wires through the knockout hole of a raceway base, fit the knockout over the nipple and fasten the nipple lock washer and bushing. If you want to restore power to the tapped circuit before completing the raceway system, cap the bare ends of the raceway wires with wire caps and electrician's tape.

2 Turning a corner. Fasten straight lengths of base raceway to the wall along the route you have chosen. Cut a base piece to fit flush with the edge of the corner and slip one end of a corner fitting into this piece; tighten the fitting screw securely to ensure a good ground between the two pieces. Slide a straight base piece over the other end of the corner fitting

and tighten the second fitting screw. Other fittings are available for inside corners and for 90° turns that route the wires up or down.

When the entire base of the system is installed, lay in the wires and secure them with wire clips. Do not at this point connect the wires to the service-panel or tapped-circuit wires.

3 Installing a receptacle. Strip ½ inch of insulation from the white and black wires—do not cut these wires—then fasten the stripped sections beneath the terminal screws of a raceway receptacle. Cut the green ground wire and connect the ground wire of the receptacle to the cut ends with a wire cap. Snap the receptacle into the raceway base with the ground wire down.

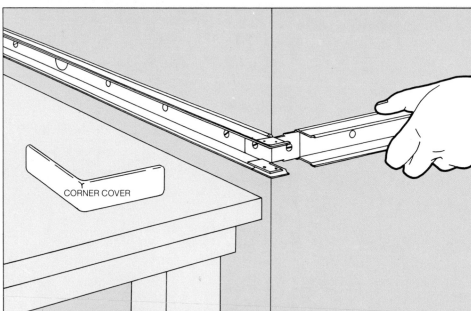

CORNER COVER

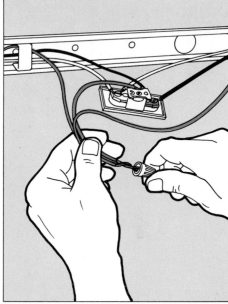

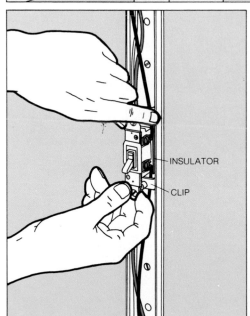

INSULATOR

CLIP

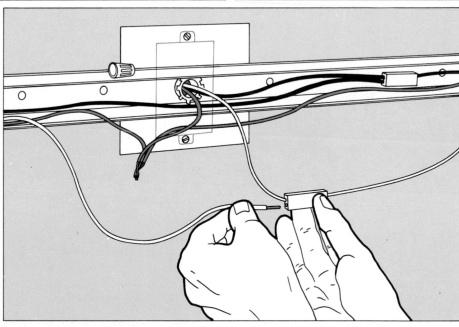

4 Installing a switch. Cut the black wire and connect the cut ends to the switch terminals; then snap into the raceway base the insulator that came with the switch, and push the switch into position. Caution: the white and green wires must run freely behind the switch, and the switch must not nick the edges of the fiber insulator when you push it into position. Screw the switch to the support clips.

5 Wiring to the power source. At an outlet box, connect the box and raceway wires, black to black and white to white, with the pressure-type connectors provided with the raceway system—to make the connection, simply strip ½ inch of insulation from the wire ends and push them into place. Stagger the location of the pressure connectors so that they do not jam the base of the raceway. Connect the box

and raceway ground wires with a wire cap. At a service panel, run the raceway wires through the offset connector and nipple and connect them to the panel (*pages 30-31*).

Test the new branch circuit (*pages 36-37*) before making the final connections. When your tests are completed, tuck all wires inside the channels and snap on the raceway covers.

Raceway for Mid-Floor Plugs

1 **Installing the base.** Run a floor-raceway base from a wall-to-floor fitting to the location you have chosen for a floor receptacle, and slide the base of a receptacle box over its end. Screw these base pieces to the floor. Lay in the wires and wire clips, then wire the receptacle and screw it to the upright supports of its base.

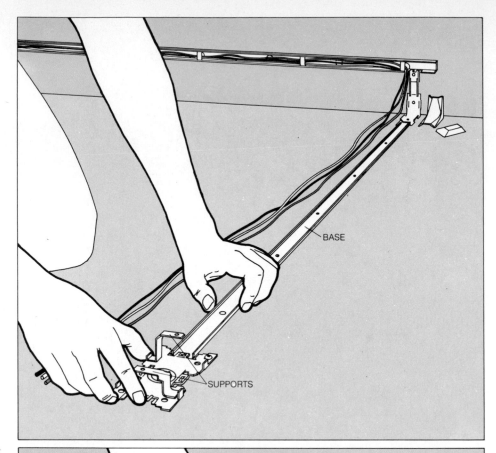

BASE

SUPPORTS

2 **Securing the cover.** Cut a length of floor-raceway cover to fit between the wall-to-floor fitting and the receptacle-box base, then force it into position by laying one edge along the raceway base and pounding the other side down with a hammer and a short piece of 2-by-4. With a pair of pliers, remove the appropriate box twist-out to make the box cover fit the end of the raceway cover, then screw the box cover to its base. Install the wall-to-floor fitting cover.

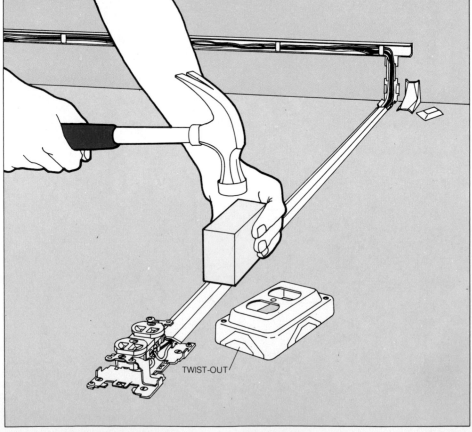

TWIST-OUT

Raceway for Baseboard Outlets

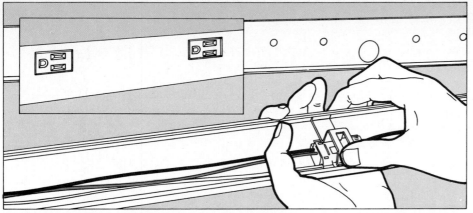

In place of a baseboard. Fasten the base of wide prewired raceway to the lower ends of wall studs along a route from which the wallboard has been cut away, then snap prewired receptacles, secured by clips, into precut holes in the raceway cover. Snap the cover pieces onto the base; the installed raceway should be nearly flush with the wallboard and should have the look of a baseboard (*inset*).

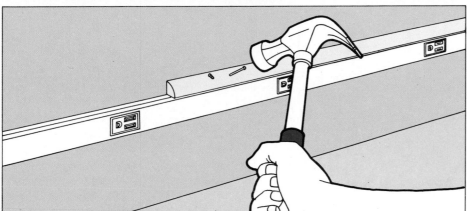

As part of a baseboard assembly. Fasten the base of narrow prewired raceway against a wall directly above a flat-topped baseboard, lay in the wires and snap on the covers, then nail quarter-round trim to the wall directly above the raceway system. For a tight assembly, glue the back of the trim and, at 16-inch intervals, drive two finishing nails close together at crossing angles to grip the wallboard and trim together.

Raceway for Quick Changes

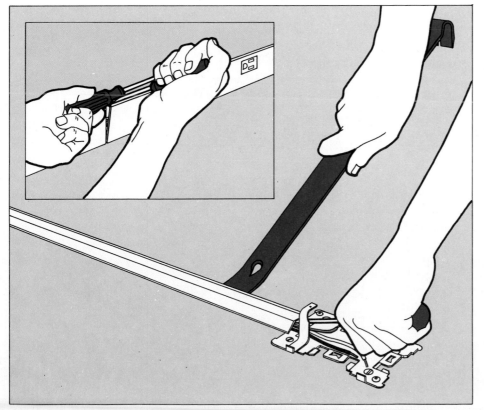

Removing raceway covers. If the manufacturer's specifications and your local electrical code indicate that you can add wires to an existing raceway, turn off power and get at the raceway base. For a floor raceway, remove the cover of a receptacle box, disconnect the receptacle and put it aside, then use a screwdriver to pry up one end of the cover while working a pry bar between the floor and an edge of the cover. Work down the raceway with the bar, lifting the pried edge at short intervals. For a wall raceway, use two screwdrivers (*inset*).

Connecting a New Circuit to a Service Panel

Connecting new cables at the service panel is the final step in installing a new branch circuit. It is also the simplest step—though anyone looking for the first time at the spaghetti-like tangle of wires inside the box might not think so.

The trick of keeping the job simple and safe is to break it into separate stages. Start by identifying the type of panel box you have and familiarizing yourself with its components. Three common types are shown at right and on pages 26-27, with their doors and safety cover plates removed. Caution: always cut off the power before removing the cover plate. Electricians usually remove fuse box pull-outs or switch main circuit breakers to "off," which shuts off power to the lower part of the panel but leaves "hot" supply cables and lugs at the top. This procedure is not recommended to the amateur. It is preferable to kill power to the panel by having the power company interrupt your service or, if your local code permits and you can confidently tackle the job, by pulling the meter yourself (page 50).

All panels contain metal assemblies called bus bars or buses, which conduct electricity within the panel, and a number of circuit breakers or fuses that protect the house wiring against short circuits and overloads. Two power buses conduct current from the "hot" wires of the 240-volt main feeder cable to the fuses or breakers, which pass it on to individual circuits. A ground/neutral bar connects the neutral wire of the feeder cable to the terminals for the neutral and ground leads of the branch circuits. The metal conductor strips that carry current in a bus assembly are generally hidden behind the fuses or breakers.

A 240-volt branch circuit draws power from both power buses, a 120-volt circuit from only one bus. It is important that the loads on the two power buses be balanced as evenly as possible; when you hook up a new branch circuit, choose a fuse or breaker position fed by the bus that bears the lighter load (to determine the amperage of the loads, use the method shown on pages 124-125).

There are two basic power-bus patterns: straight bus and split bus. In a panel with a straight bus (opposite and page 27), one main breaker switch or fuse pull-out controls the power to all the breakers or fuses. In a panel with a split bus (page 26), current flows through divided buses, and more than one switch or pull-out must be used to cut off all the power in a house. Adding a new 240-volt branch circuit may be impossible if you have a straight-bus panel fitted with fuses: many such panels have two or three 240-volt positions, and every available position may already be filled. In this case, either add a subpanel (pages 32-35) or replace the main panel with a larger circuit-breaker panel (pages 40-44).

A fuse box generally contains two different types of fuses: cartridge and plug. Cartridge fuses, installed in pull-outs that resemble small drawers, protect 240-volt circuits carrying large currents; plug fuses, which screw into sockets, protect 120-volt circuits carrying 30 amperes or less. Both types work in the same way: when a circuit is overloaded, a metal strip in the fuse melts, cutting off power to the circuit. When the cause of the overload has been found and corrected a new fuse must be installed.

In a circuit-breaker panel, single-pole breakers, making contact with a single power bus, protect the 120-volt branch circuits; double-pole breakers, spanning both buses, serve the 240-volt circuits. An overload will cause a breaker to switch the power off; when the overload has been corrected the breaker can be reset and switched back to "on."

Knockout disks on the sides, top and bottom of the panel box must be removed to make holes for branch-circuit cable clamps. Draw the cables far enough into the box to provide slack for tucking the leads out of the way. Outside the box, label each cable with a marked piece of masking tape indicating its amperage, voltage and function. Inside the box, strip off the outer sheaths of the cables. Caution: if you are working with existing aluminum wiring, use anticorrosive paste, lugs and bus bars approved for use with aluminum wires.

Branch circuits, of course, can only be attached to a panel that has vacant positions for breakers or fuses and a total amperage that can handle the additional load (pages 124-125). If you have a fuse panel with no vacant positions, install a subpanel or replace the main one. In a fully wired breaker panel, "piggyback" or "skinny" breakers can be used to make room for two branch circuits in one bus opening. A piggyback—technically called a tandem breaker—is the same size as a normal breaker; it contains two circuit breakers in a single plastic case. A skinny—technically called a half-module breaker—resembles a normal breaker but is only half as wide so that two can fit in a normal opening. If there are extra spaces and extra electrical capacity in the panel, hook the circuit to a new breaker as you would in a subpanel (page 34).

Branch-circuit devices produced by different manufacturers are generally interchangeable—service-panel equipment rarely is. When purchasing new circuit breakers, note the make and model number of your service panel to make sure that the new breakers you buy will fit. For a fuse panel, you also will need fuse reducers; the electrical codes in most localities now require that these small, permanent devices be used to prevent high-amperage fuses from being used in low-amperage circuits.

Canadian service panels, manufactured to meet the requirements of the Canadian Electrical Code, have a few distinctive features. A separate compartment at the top of the box contains the fused main shutoff switch, main breaker or pull-out. Branch-circuit cables cannot be routed through this separate compartment; instead, they enter the service panel through the sides or bottom. The main ground cable, which does pass through the compartment, must be an insulated wire or taped if it is smaller than No. 8. In Canadian panels, also, the bare or green ground lead of a branch circuit is attached to a separate ground bus bar, the white neutral lead is attached to a neutral bus bar, and the black, or black and red, leads are wired as shown here.

Do not restore power to the box until you have tested the new circuits and their outlets (pages 36-37). Large appliances, such as pumps, freezers, heating units and air conditioners should be unplugged, then plugged in one by one, so that they do not overload the system by starting simultaneously when the power is turned back on.

The Three Basic Panels

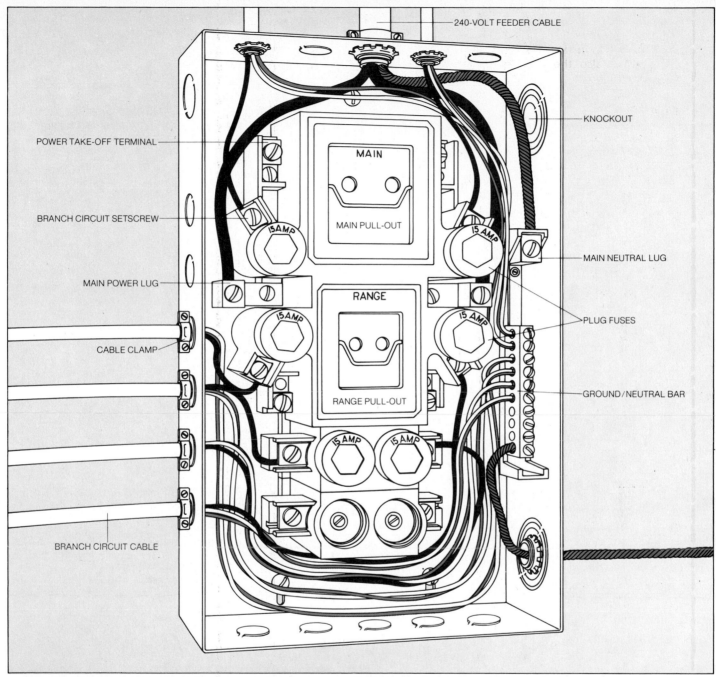

240-VOLT FEEDER CABLE

KNOCKOUT

POWER TAKE-OFF TERMINAL

BRANCH CIRCUIT SETSCREW

MAIN NEUTRAL LUG

MAIN
MAIN PULL-OUT

MAIN POWER LUG

RANGE
RANGE PULL-OUT

PLUG FUSES

CABLE CLAMP

GROUND/NEUTRAL BAR

BRANCH CIRCUIT CABLE

A straight-bus fuse panel. A 240-volt cable enters this panel box at the top. The two power leads of the cable are attached to the main power lugs, which are linked to the fuse receptacles by two 120-volt power buses—vertical metal bars, hidden behind the devices in the center of the panel. The braided neutral lead goes to the main neutral lug, linked to the branch-circuit neutral and ground wires by the ground/neutral bar. A single pull-out can be removed to turn off all the power; a smaller pull-out beneath protects a 120/240-volt branch circuit that feeds a large appliance such as an electric range, water heater or air conditioner. Plug fuses ranging up to 30 amperes protect six 120-volt circuits. The fuses at right conduct current from one bus; the fuses at left, from the other. Two power take-off terminals at the top of the panel could provide wiring positions for a subpanel (page 32). The branch-circuit cables enter the panel through knockouts and are fastened to the box with cable clamps. The branch-circuit neutral and ground leads are wired to the setscrews of the ground/neutral bar, while the black hot leads are attached to the branch-circuit setscrews next to the fuses.

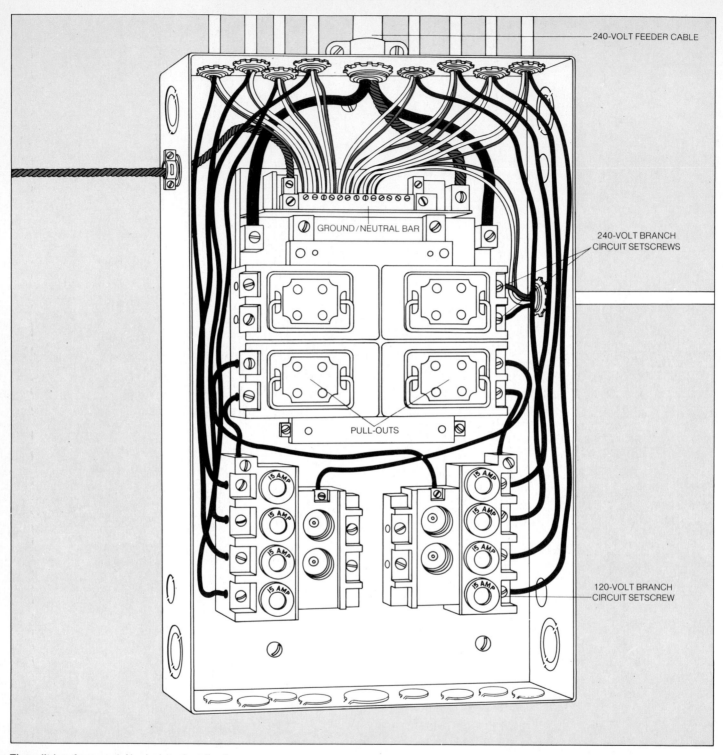

240-VOLT FEEDER CABLE

GROUND/NEUTRAL BAR

240-VOLT BRANCH
CIRCUIT SETSCREWS

PULL-OUTS

15 AMP
15 AMP
15 AMP
15 AMP

15 AMP
15 AMP
15 AMP
15 AMP

120-VOLT BRANCH
CIRCUIT SETSCREW

The split-bus fuse panel. No single main pull-out in this panel will shut all the power off. The service cable goes to two power buses (not shown), split into sections, each connected to an individual pull-out. In this example, four pull-outs must be removed to shut off all the power. Plug fuses for the 120-volt branch circuits are protected by two pull-outs. The remaining pull-outs serve 240-volt circuits for such appliances as the water heater, range or air conditioner.

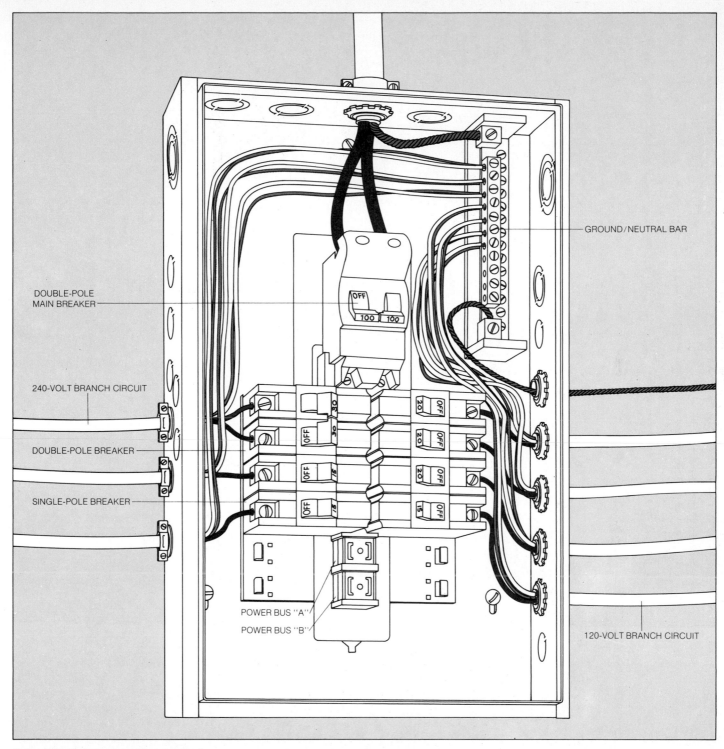

GROUND/NEUTRAL BAR

DOUBLE-POLE
MAIN BREAKER

240-VOLT BRANCH CIRCUIT

DOUBLE-POLE BREAKER

SINGLE-POLE BREAKER

POWER BUS "A"

POWER BUS "B"

120-VOLT BRANCH CIRCUIT

A circuit-breaker panel. In this example of a straight-bus breaker panel, the service cable feeds a 100-ampere double-pole main breaker at the top of the panel. This main breaker contains two internally linked switches—one for each of two 120-volt power buses—that can cut off the power to the entire house. The power buses pass current to breakers of smaller amperages. (In a split-bus breaker panel, a relatively uncommon design similar to the fuse box at left,

as many as six separate high-amperage circuit breakers would have to be switched off to interrupt power throughout the house.)

Single-pole circuit breakers that protect the 120-volt branch circuits are connected alternately to one power bus or the other in an A-B-A-B pattern. Double-pole circuit breakers, each connected to both the A and B buses, protect the 240-volt branch circuits.

The Basic Safety Test

Checking the panel. Before working on a service panel, switch off power at the main fuses or circuit breakers, then disconnect the panel entirely (page 50), calling the power company if necessary. Ground one prong of a voltage tester on the ground/neutral bar and touch the other prong firmly to each of the hot supply lugs near the main shutoff. If the tester registers any indication of voltage, stop work immediately and call an electrician. As insurance, also test the branch circuits (below): in a fuse box (left), test each of the branch-circuit setscrews; in a breaker panel (right), test the two power buses in the center of the panel, snapping out adjoining single-pole breakers to reveal the power buses if necessary.

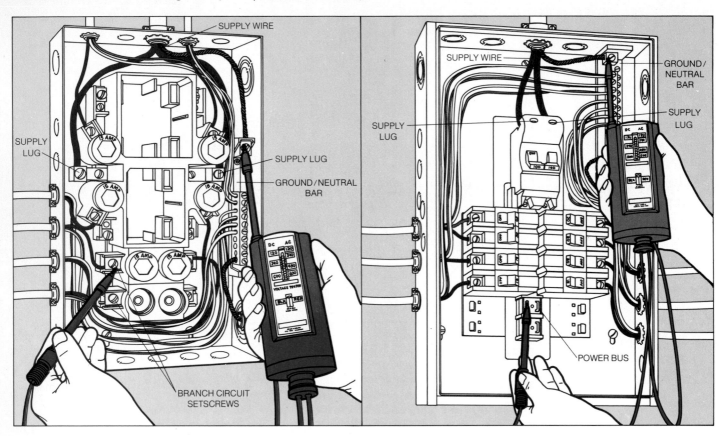

Removing a Panel Knockout

1 **Punching out the center.** Outside the box, place the tip of a nail set on the central section of a multisection knockout opposite the tie to the first knockout ring. Tap the knockout center inward, then grip it from inside the box with pliers and work it free. Removing the center section is sufficient for cables smaller than No. 10. For larger cables, remove one or more knockout rings as described in Steps 2 and 3.

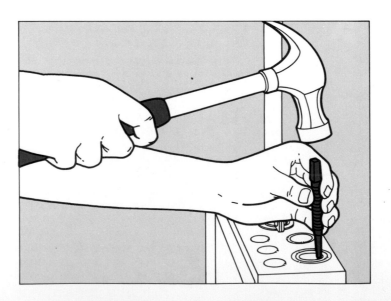

2 Prying up a ring. Pry up both sides of the knockout ring with the tip of a screwdriver set midway between the ring's ties. To make this step easier, use pliers set flat against the box to lever the screwdriver.

3 Removing the ring. Grasp both sides of the knockout ring with pliers and work the ring free. To enlarge the hole further, remove additional rings by prying them in or out with a screwdriver, then working them free with pliers.

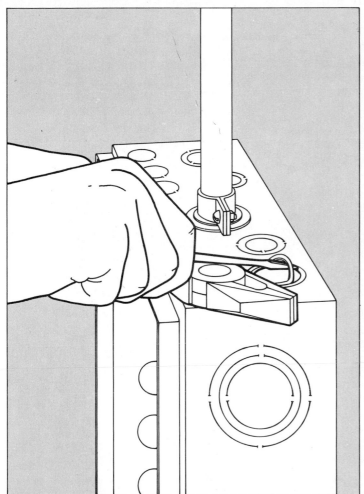

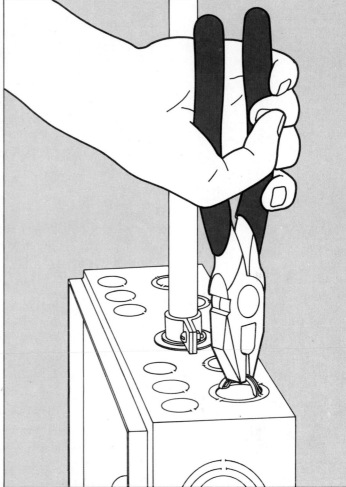

Closing a Knockout Hole

Using a filler plate. According to code regulations, an unused knockout hole must be sealed. Place a knockout filler plate, sold for the purpose, into the hole and tap it into place with a block of wood and a hammer.

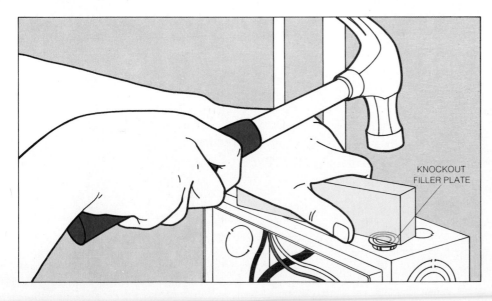

KNOCKOUT FILLER PLATE

Wiring a Circuit to a Fuse Panel

1 Making the connections. With power to the service panel off, run the new circuit cable through a knockout hole and fasten it to the panel with a cable clamp. For a 120-volt circuit, attach the white neutral lead to the ground/neutral bar, the green or bare copper ground lead to the bar, and the black lead to an unused branch-circuit setscrew. For a 120/240-volt circuit, attach the red and black leads to the two setscrews of an unused pull-out. Caution: never attach these leads to two 120-volt setscrews.

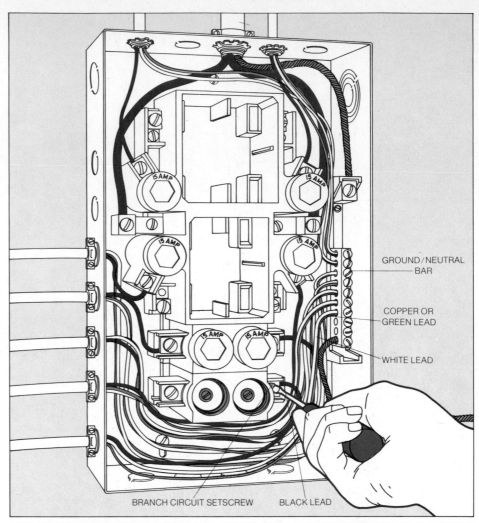

GROUND/NEUTRAL BAR

COPPER OR GREEN LEAD

WHITE LEAD

BRANCH CIRCUIT SETSCREW   BLACK LEAD

2 Installing a fuse adapter. To be sure that you will protect the new branch circuit with a fuse of the correct rating, screw a fuse adapter with the same amperage as the circuit into the fuse receptacle. Caution: check the ampere rating of the adapter carefully before you screw it in; the device has a spring-loaded barb that prevents it from being removed after installation.

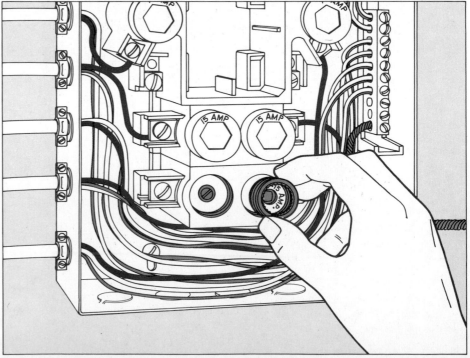

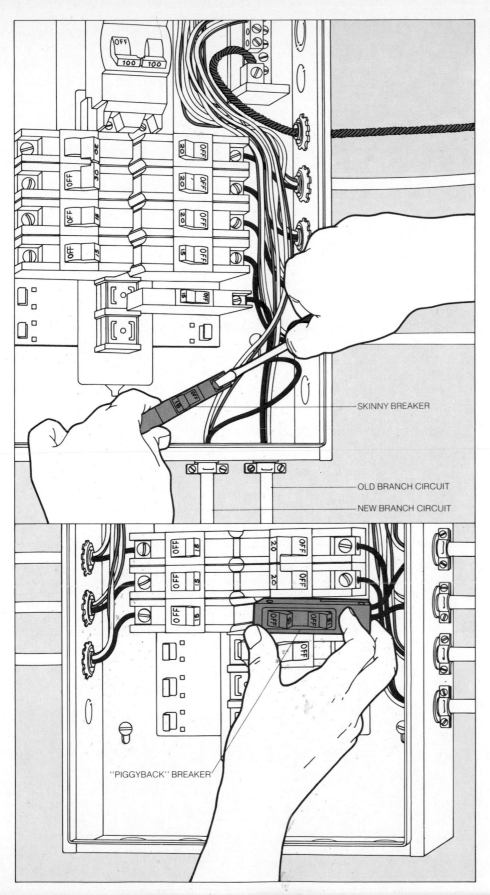

Wiring a Circuit to a Spacesaving Breaker

Skinnies and piggybacks. Directions for installing a conventional circuit breaker are given on page 34. If there is no room in a box for such a breaker, you can replace an existing one in a 120-volt circuit with either a pair of skinnies (*left*) or a piggyback (*below*) of amperage the same as the removed breaker. With power to the panel off, attach the neutral and ground wires of the new circuit to the ground/neutral bar in the box. Then switch a single-pole breaker to "off," remove it from the service panel and unhook the black lead attached to it. Wire this lead to a skinny breaker or to one of the terminals at the base of a piggyback. Attach the black lead of the new branch circuit to a second skinny or the second terminal of a piggyback.

Caution: never wire the hot leads of a 240-volt branch circuit to a piggyback or a pair of skinnies. These breakers fill a position on only one power bus and must be restricted to 120-volt circuits.

SKINNY BREAKER

OLD BRANCH CIRCUIT

NEW BRANCH CIRCUIT

"PIGGYBACK" BREAKER

The Flexibility of a Subpanel

When you want to add new branch circuits to your electrical system, it is sometimes best to feed them through a new subpanel—a scaled-down version of a service panel—instead of connecting them directly to your existing service panel. A subpanel can be in a remote location to make wiring distant branch circuits easier. Instead of snaking every branch circuit back to the main panel, you need to run only one high-capacity feeder cable between the main panel and the subpanel. But the principal use of a subpanel is to add circuits to a main panel that is equipped with fuses rather than circuit breakers. A fuse box rarely has room for additional fuses. The solution is a new subpanel mounted near the old fuse box to protect the added circuits.

Before you install any subpanel, calculate the total load it will have to carry and make sure that the main panel would not be overloaded by the new circuits you add; if it would be, you should probably have the electric company increase the service to your house so you can install a new service panel (pages 40-44). Determine the size of the subpanel and the feeder cable by calculating the amperage of the branch circuits you plan to connect using the method shown on pages 124-125, and then consulting the chart of amperages and wire sizes there.

A typical subpanel is fed with 240 volts from the main panel by a three-wire cable and contains breaker positions for two to six branch circuits. If all you have is 120-volt service at the main entrance panel, you may be able to add a subpanel to it, but it is generally preferable to replace a 120-volt main panel with a 120/240-volt circuit-breaker panel of ample capacity (in some localities, codes prohibit the addition of subpanels to 120-volt main panels).

Before installing the subpanel, run the branch-circuit cables and the feeder cable to the subpanel location, arranging them so that all enter the box from one side. Then connect the cables to the subpanel terminals and test the installation; in some areas, have it approved by an electrical inspector before you attach the feeder to the main panel. Although professional electricians sometimes merely switch off the main fuses or breakers before connecting the feeder cable, for safety amateurs should cut power entirely. Have the power company disconnect service to your house or, it if is allowed in your locality, pull out your meter (page 50) before connecting the cable.

1 Mounting the ground bus bar. Holding the ground bus bar in the subpanel box where it will not interfere with the knockouts you will use, drill through the mounting hole in the bar and through the back of the box. Screw the bar to the box. Open knockouts along one side of the box (pages 28-29). For the flush-mounted box shown here, pull the branch-circuit and feeder cables through separate knockouts and secure the cables with cable clamps.

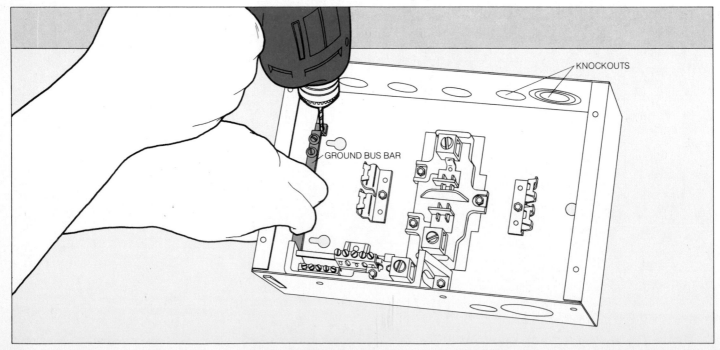

KNOCKOUTS

GROUND BUS BAR

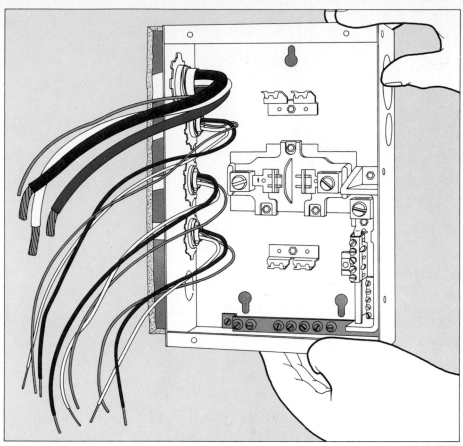

2 Mounting the box. Push the subpanel box into the wall, wired side first, so that its front edges are flush with the surface of the wall. Then screw the mounting side of the box flush against the stud with 1-inch round-head wood screws.

If the subpanel is to be surface-mounted on an unfinished wall mount it first *(page 42)*; then pull in and secure the wires.

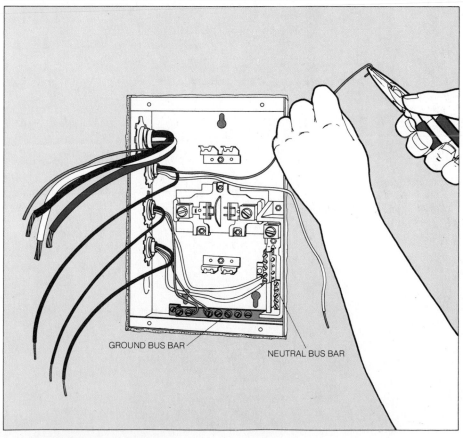

GROUND BUS BAR

NEUTRAL BUS BAR

3 Connecting the branch circuits. After testing the unconnected branch circuits *(pages 36-37)*, connect each green or copper ground wire to the ground bus bar. Connect each white neutral wire to the neutral bus bar.

4 **Wiring the circuit breakers.** Push each 120-volt branch-circuit black lead into the terminal at the bottom front of a single-pole circuit breaker of the proper amperage and tighten the set-screw. Slip the breaker over its guide hook and snap the breaker down to lock it into place.

If the branch circuit is 240 volts, use a double-pole circuit breaker and insert both the black and red leads into its terminals.

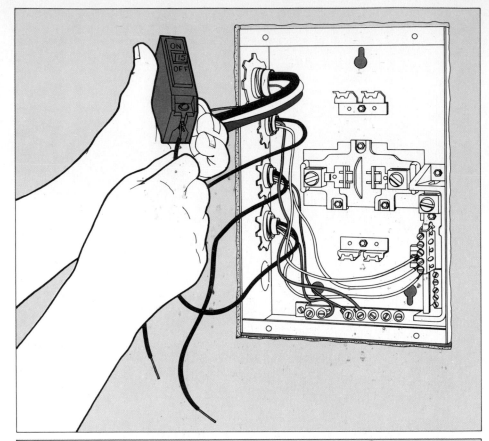

5 **Connecting the feeder cable.** Insert the stripped end of the white feeder wire into the neutral lug at one end of the neutral bus bar and tighten the setscrew. Attach the bare copper or green insulated ground wire to the ground bus bar as in Step 3, then attach the black and red conductors to the subpanel power terminals.

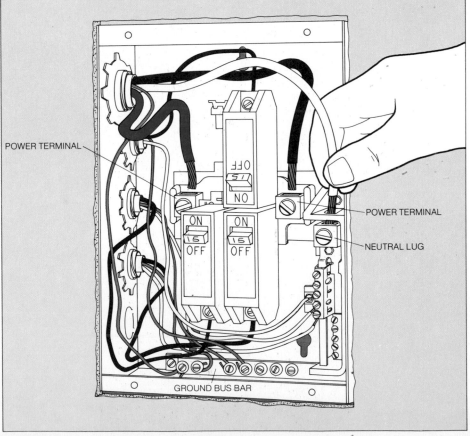

POWER TERMINAL

POWER TERMINAL

NEUTRAL LUG

GROUND BUS BAR

6 **Wiring the feeder to the service panel.** Connections are simpler with a circuit-breaker main panel than with a fuse box (*right and below*). In either case, first have the power company turn off all current, or, if you are qualified, pull the meter yourself; check that the panel is dead. If it is a circuit-breaker type, connect the feeder as you would a 240-volt branch circuit (*page 44, Step 9*). In a fuse box, pull the fuse blocks and attach the feeder neutral conductor to the neutral lug (*right*). Attach the ground wire to the branch-circuit ground bus bar.

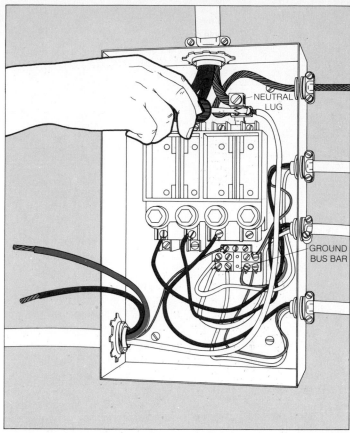

7 **Connecting power leads to a fuse box.** Insert red and black feeder leads into the take-off terminals and tighten the setscrews. After making all connections to either a fuse box or circuit-breaker panel, replace the safety cover on the subpanel. Then restore power and test the new branch circuits as described on page 36.

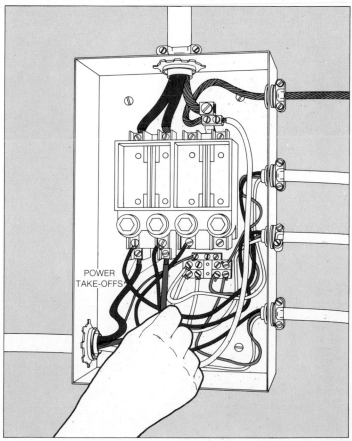

Testing and Troubleshooting New Installations

Nothing is more frustrating than mistakes made in wiring a circuit; locating flaws can be uncertain, mystifying and—occasionally—dangerous. Nevertheless, electricians have worked out a series of procedures that usually finds the trouble quickly and safely.

Almost always, the trouble takes one of two forms: short circuits that totally disable the new circuit, and wiring mistakes that make the circuit unsafe or prevent it from functioning properly.

A short circuit is a fault that causes a direct connection from a hot wire to either a neutral wire or the ground (and in 240-volt circuits from one hot wire to the other). Test for shorts before you connect the circuit at the service panel. Separate and tape the stripped ends of hot and neutral wires at the panel and turn all switches to "on"; do not screw light bulbs into sockets or plug appliances into receptacles. Then go to the box at the end of the run and hook a continuity tester across all possible combinations of hot wires, connecting each to other hot wires, to neutral wires and to ground. If the bulb glows, the circuit is shorted. Also test each neutral wire to ground; a glow indicates a grounded neutral wire that must be repaired. To find the cause of trouble, look for the clues shown below—which, incidentally, also serve for troubleshooting short circuits in existing wiring (box, opposite). When you have repaired the wiring fault, you can connect the circuit at the panel with the assurance that it will not blow fuses or trip circuit breakers, and can go on to test the circuit for mistakes that can be discovered only with the current on.

Normally, these mistakes will be at a receptacle, and can be diagnosed with two inexpensive hand tools: a receptacle analyzer and a meter that reads common house-wiring voltages on a moving bar mounted in a small window (opposite, top). Of the five mistakes that can be detected by the analyzer (and indicated on its panel), one deserves particular concern. If hot and neutral wires are reversed at the receptacle terminals, a serious hazard is introduced, for polarized plugs (three-pronged plugs and plugs with two prongs of different widths) may send current to the metal cabinets of such appliances as tape decks, radios, refrigerators and television sets.

Two other kinds of miswiring detectable by the analyzer can energize the outlet box in which a receptacle is mounted: reversed hot and ground wires, and a combination of an open—that is, disconnected—ground wire and a hot wire shorted to the box.

Though these tests are made at outlet boxes alone, they are likely to answer your purposes: electricians estimate that nine out of 10 wiring defects are in the wiring of boxes rather than in runs of cable. But finding problems in circuits can be tricky, and a mistake in diagnosis very serious. Even if you locate one wiring error there may be another; always retest. Unless you are absolutely certain you have located all of the defects, and always if you suspect the defect is in a run of cable, call an electrician.

Three Common Causes of Shorts

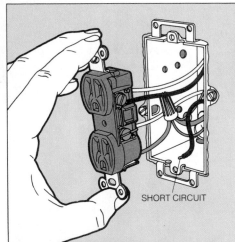

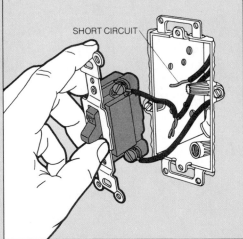

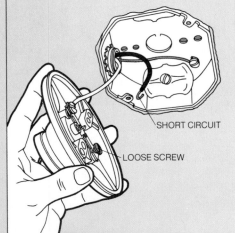

The basic clues. One by one, pull receptacles, switches and fixtures out of the outlet boxes and look for hot wires that have broken free from their terminals and sprung into contact with the box or a ground wire. Three common causes are broken wires, nicked by wire strippers set a size too small (above, left); downward-pointing wire caps that fall off and let a wire spring away (center); and loose terminal screws (right). Repair the short circuit by reconnecting the sprung wire. If a broken wire is too short to reach its terminal, either loosen the cable clamp and pull in some slack, or wire-cap the short wire to an extender wire about 2 inches long.

If you find no sprung wires, look for overtightened cable clamps that have cut through cable sheathing and wire insulation, or for hot wires that you have accidentally bared when you stripped off the cable sheathing. You may be able to make a repair by pulling slack cable into the box; if not, replace the run of cable.

When you have made your repairs, screw all receptacles, switches or fixtures back into place and run another continuity test. When you are certain that all of the shorts have been eliminated, install cover plates on the outlet boxes, screw the light bulbs into the fixtures and connect the circuit at the service panel.

Reading a Receptacle

Analyzing a 120-volt circuit. To check receptacles, plug a receptacle analyzer into each receptacle in the circuit. If its coded lights do not signal "ok," turn off the circuit and repair the wiring defect that is indicated on the analyzer panel. Correct any reversed wiring by switching the indicated wires; find open connections in the same way you would a short (*opposite*).

To check a switch that fails to work, turn off the circuit. Repair any open connections; then turn the switch to "on" and run a continuity test across the switch terminals. If the tester bulb does not glow, replace the switch.

Testing a 120/240-volt circuit. Insert the probes of a bar-meter voltage tester into the straight slots of the receptacle; the reading should be 240 volts. Now insert one probe into the L-shaped slot and the other into each of the straight slots in turn; the readings should be 120 volts in both cases. If any of the three readings is incorrect, turn off the current and rewire the receptacle with the white wire in the terminal marked "white" and the red and black wires in the other two terminals.

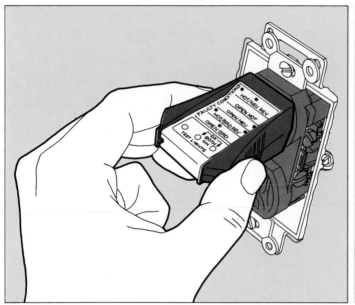 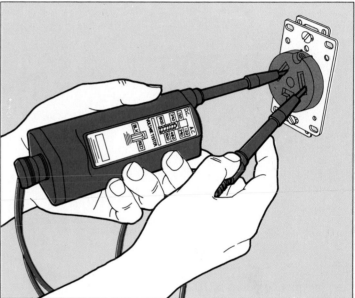

What to Do When Fuses Keep Blowing

When a fuse blows or a breaker trips, the fault is either an overload or a short circuit. Check for overloading, the most common cause, by adding up the amperages on the circuit, allowing for the starting draw of motors and fluorescent lights (*pages 124-125*). Correct overloading temporarily by unplugging a few of the appliances, and permanently by installing a new branch circuit.

In a circuit that is not overloaded, search for a short. Unplug appliances from all dead outlets and turn switches that control inoperative fixtures to "on." Replace the fuse or reset the breaker. If the fuse does not blow or the breaker trip, the short is in one appliance. You may find it by inspecting the appliance plugs for strands of wire between the prongs, and their cords for frayed or bared spots. Failing that, have an appliance service shop make any needed repairs in the appliance.

Shorts within the circuit itself are rare but possible. Wires heat up slightly when current flows through them, then cool off when the current stops; the corresponding cycle of expansion and contraction can dry and crack the insulation or break a nicked wire. Elsewhere, wires can even break from constant slight bending or vibration.

If you have a main disconnect switch or another safe way of turning off the power to the entrance panel, you can unhook an existing circuit at the panel and troubleshoot it for shorts as you would a new circuit (*opposite*). Missing insulation or a black burn spot where a hot wire has sparked against a grounded outlet box may help call your attention to a short. Even if you find a clear case of a short, search the boxes on the circuit for others. If you cannot be sure that you have found and repaired the defect, or if it appears to be in the cables, have an electrician come in and find the trouble.

2

Bringing More Current to a House

The meter that measures the flow of power into your house, the service panel that distributes the power and the cable that links the two—all are designed to carry a specific amount of current. The amount (it may be 60 amperes, or 100, or as much as 200) is indicated on the main fuse or circuit breaker, and it is generally sufficient for the needs of the house. But from time to time you will find that new circuits or large appliances demand more current than your meter, panel and cable provide. To get it, you must put in higher-amperage components that can draw more current from the power company's lines. In the process you will have opportunities to improve your electrical system in other ways—by replacing an unsightly overhead service with an underground one, for example; by moving the meter to a less conspicuous location; by adding a main-disconnect switch, which enables you to shut off all power to the house whenever you wish, rather than calling the power company or having an electrician pull the meter; or by installing a lightning surge arrester to protect your house from high-voltage surges in the company's lines.

The power company will usually plan much of the main job for you. Its service representative will visit your house to discuss your plans and show you where to mount the new meter and service panel; if you install an underground service he will usually map routes for the trenches. Most companies provide the meter box and the cable, either free of charge or for a fee. A few will even do part of the work for you—installing the meter box, for example, or digging trenches for underground cable. All are responsible for the final connection to their own power lines.

Your next step is to obtain an electrical construction permit. You will need to prepare a rough sketch of your entire project, showing the locations of all the components, the routes of new circuits, and the sizes of loads, cables and fuses or circuit breakers. In some localities you may have to take a quick test—usually about a dozen short-answer questions—on the basic safety requirements of your local electrical code, available at the electrical inspection office.

When your project nears completion, arrange an appointment for an electrical inspection and for a power company crew to make the final connection. The electrical inspector must be sure that boxes, fixtures and cables are properly mounted and protected; that fuses (or circuit breakers) and cables match the amperage of their loads; and that hot, neutral and ground wires are properly connected. Once your work has been approved, power company linemen will make a quick inspection of their own and then will disconnect the cable that supplied your old meter, connect a new cable and plug in a higher-amperage meter, energizing your new service.

Inside: Replacing the Main Service Panel

Occasionally, a homeowner may put in a new service panel for reasons of convenience—to replace fuses with circuit breakers, perhaps, or to provide wiring connections for additional branch circuits. More often, a new panel upgrades an electrical service to handle more current safely and meet the high electrical demands of the modern home.

If your main fuse blows from time to time, or your main circuit breaker trips, you probably need more current; if you plan to install branch circuits that would overload an existing panel, you certainly do. To get more current, you must install not only a new service panel of greater amperage but also a meter socket *(pages 51-53)* to match, a new weatherhead or underground connection *(pages 58-63)* and the cable, conduit and wire that link these service components.

Use the formulas on pages 124-125 to calculate the current you need so that you will know the capacity required of the new service panel. Panels come in capacities of 100, 125, 150 and 200 amperes; though the National Electrical Code requires a panel of no greater capacity than your total electrical load, it is a good idea to get a model one size larger than this minimum to allow for additional demand in the future.

In each amperage rating, panels come in several sizes, depending on the number of single-pole circuit-breaker openings they contain. To choose one for yourself, count the circuits you already have (if you have any piggyback breakers in the panel, count each as two circuits). Then add at least six additional openings.

At the same time, decide whether you want a panel with split-power buses *(page 26)* or a more expensive and versatile straight bus; whether you want a weatherproof case or one restricted to indoor use; and whether you want a set of built-in main circuit breakers or a set of lugs for a separate main disconnect *(pages 48-49)*. Your local code may require special variations—either a separate box housing a meter and a main disconnect, instead of a disconnect located at the top of the panel; or a single box

housing both a meter and a service panel. In these combinations, the manufacturer makes the connections between the meter and the disconnect or panel; all other wiring procedures are the same as those on the following pages.

Finally, you will need a cover for your service panel. Covers usually are sold separately and are available for two types of mounting: flush mounts, in which the panel fits within a wall and is attached to two studs; and surface mounts, especially suitable for masonry walls, in which the panel is screwed to a plywood backing fastened to the wall.

Along with your new panel you must buy the supply wiring that will connect it to a new meter. This wiring usually is encased in a service entrance concentric (SEC) cable; if your local code requires that supply wires be protected with conduit, use individual, insulated wires, which are cheaper and easier to work with. The size of the supply conductors depends on the amperage of your new service; consult the charts on pages 124-125 to determine the correct gauge. If you use aluminum wire, make sure the lugs of the new panel are approved for use with aluminum—they should be marked AL or AL/CU—and coat the bare ends of the wires with anticorrosive paste before making connections.

The service panel is always the first part of a new service to be installed. Always be sure that the supply of power to your house is stopped *(page 50)* so that you can work on it in safety. After you have mounted the new panel and wired its branch circuits and supply cable, install the new meter and the weatherhead or underground service, along with the wiring that connects them. When your work has been inspected and approved, the electric company will make the final connections to restore service.

Unless your house has a separate main disconnect, you must mount your panel as close as possible to the point where supply wires from the new meter will enter the house. Access to a panel must not be obstructed; allow at least 3 feet of clear space in front and on either side of

the panel and do not place it in a confined or wet location such as a closet or a bathroom. All the branch circuits must be labeled according to their function—"washing machine," "range," "general lighting" and so on.

These basic rules are easy to follow, but a Code provision forbidding splices within a panel box may cause some extra work. Electrical inspectors usually permit one or two splices, but if a number of the old panel's branch-circuit wires are too short to allow for a service change, you may have to extend them with new runs of cable, connected to the old ones in junction boxes mounted next to the new panel. If all the wires for your branch circuits are too short, install the new service panel next to the old one and use the old one as a junction box: remove its chassis, connect new lengths of cable to the old wires with wire caps and run the cables through unused knockouts to the new panel. Replace the old panel's cover and door with a solid sheet-metal cover.

The Code also requires that you correct any faulty wiring in the old panel. For example, the amperage of a breaker must not exceed the amperage of its branch-circuit wires—15 amperes for No. 14 wire, 20 amperes for No. 12 wire and so on. Use the chart on page 124 to determine the amperage of each hot circuit wire; do not rely on the amperage of the fuses or breakers in your old box for this information. To prevent one power bus from becoming overloaded while the other is underused, divide your wattage load evenly between the two. And before you call for an electrical inspection, make sure your grounding cable meets the requirements described on pages 45-46.

While you are working on the service panel, you may want to install two optional safety devices. A surge arrester *(page 47)* protects your house from fire and your electrical system from damage when lightning strikes nearby power lines. A separate main circuit breaker or disconnect *(pages 48-49)* enables firemen to shut off power easily in an emergency—or permits you to shut off the power whenever you choose.

1 **Marking branch circuits.** With power cut off to the panel, write the function of each 120-volt circuit on a piece of tape and wrap the tape around the circuit's black, or hot, wire. A 240-volt circuit normally has both black and red hot wires; tape each pair together when identifying it. A white wire connected to a fuse is a hot wire for a balanced 240-volt circuit; mark it red with paint or tape, find the black wire it is paired with by tracing the wire back to its cable and tape the two together. If the circuit serves several fixtures mark it "general lighting."

2 **Disconnecting the wires.** Using narrow-jawed diagonal-cutting pliers, sever wires wrapped around screw terminals; make the cuts as close to the terminals as possible. At the set-screw pressure connectors called lugs, loosen the setscrews and pull the wires out. Unscrew the lock nuts fastened around cables immediately inside the box and slide the nuts off the wires; loosen the connector screws that secure the cables to the outside of the box. Remove all cable staples or clamps near the box, taking care not to damage the cable sheathing.

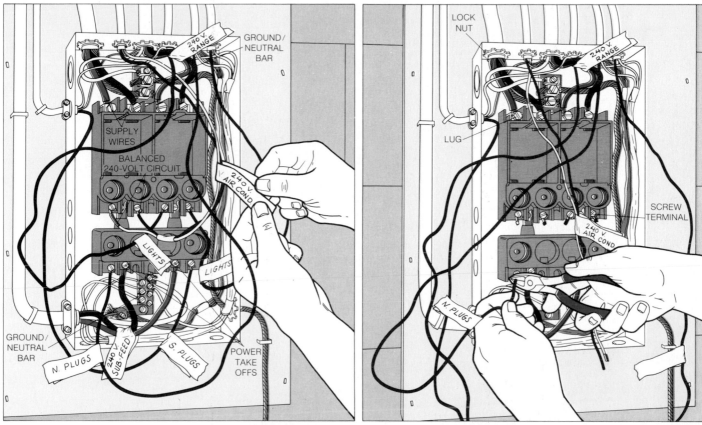

3 **Removing the old panel.** Remove the screws that fasten the panel to the wall and, while a helper eases the panel down, feed the wires out through the knockouts. If you plan to surface-mount the new panel, make sure that the old backing extends 8 inches beyond the new box on all sides. If it does not, replace it with a larger sheet of ¾-inch plywood.

If you plan to mount the new panel flush inside a framed wall, cut a hole through the wall surface between two studs. The hole should be 14½ inches wide and as high as the new panel.

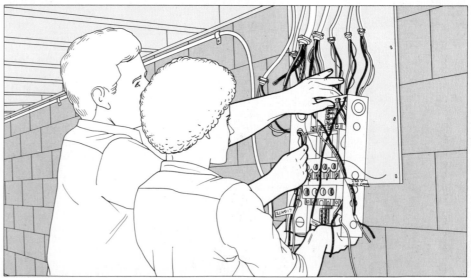

4 **Bonding the new panel.** If you are installing a separate main disconnect (*pages 48-49*), you can mount the panel directly as described in Step 5; otherwise you must first bond the panel box to the ground/neutral block with a screw or clip (*right*). If your box has a bonding screw, thread it into the hole predrilled in the box behind the block. If your panel has a sheet-metal bonding clip (*inset*) screw the broad end of the clip to the box, bend the clip into a U shape with long-nose pliers and insert the narrow end into a lug in the ground/neutral block.

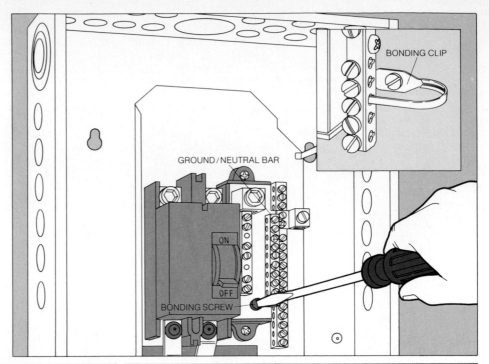

BONDING CLIP

GROUND/NEUTRAL BAR

BONDING SCREW

ON

OFF

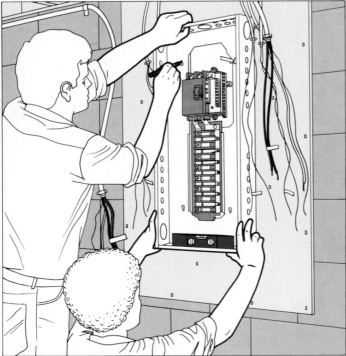

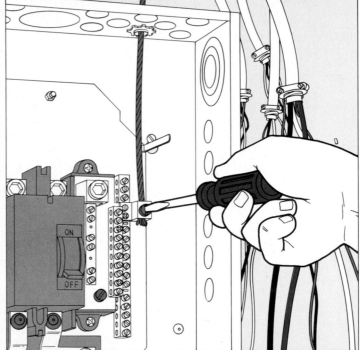

5 **Mounting the panel.** To mount the panel on plywood backing (*above*), have a helper hold it level and centered on the plywood while you mark the location of the top screw holes. Remove the panel, run round-headed wood screws partway into the plywood, slip the panel over the screws and tighten them. Make sure that the box is level, then install the bottom screws.

To flush-mount a panel adapt the method shown on page 33, Step 2, screwing the panel to the studs on both sides.

6 **Grounding the panel.** Unless you are installing a separate main disconnect (*pages 48-49*) or a combination meter-main disconnect, you must ground the box as illustrated above. Attach a bare stranded copper wire to the lug at the side of the ground/neutral block and run it through a knockout to a grounding electrode connection (*pages 45-46*). If you use the grounding wire from your old panel make sure it is large enough for the new service: No. 8 copper for 100 or 125 amperes, No. 6 for 150 amperes and No. 4 (No. 3 in Canada) for 200 amperes.

7 Connecting the supply wires. How you make the main connections depends on the direction from which cable enters the box and the kind of cable used. If you have a separate main disconnect, the cable will be service entrance round (SER) with four wires (*below, right*), otherwise the conductors will be encased in three-wire service entrance concentric (SEC).

If SEC cable enters above the panel (*left*), pull 18 inches through the large knockout in the top of the box, then strip the cable sheathing inside the box and twist the neutral wires together (*page 15*). To fit the wires to their lugs, cut through them with a hacksaw or sever a few strands at a time with diagonal-cutting pliers. Strip 1 inch of insulation from the ends of the hot wires and work the hot and neutral wires into their lugs. Tighten the lugs with a hex wrench and secure the cable clamp at the top of the panel.

If the supply wires enter below the service panel (*center*), it probably will be encased in conduit. Secure the conduit to the large knockout at the panel bottom and draw the wires into the box. Run the hot wires up the left side of the panel to their lugs and the neutral up the right side to its lug. Caution: at the top of the panel, keep the neutral well away from the hot wires and lugs to eliminate the possibility of arcing.

In a system with a separate main disconnect (*right*), install a separate grounding bar in the pre-drilled holes at the right side of the box. Strip the sheathing from the SER cable and the insulation from the ends of the two black hot wires and the gray neutral wire. Connect these wires to their lugs. Bend the fourth cable wire, a bare grounding conductor, away from the neutral lug; run this wire forward to the upper right-hand corner of the box, then down to the large lug on the grounding bar.

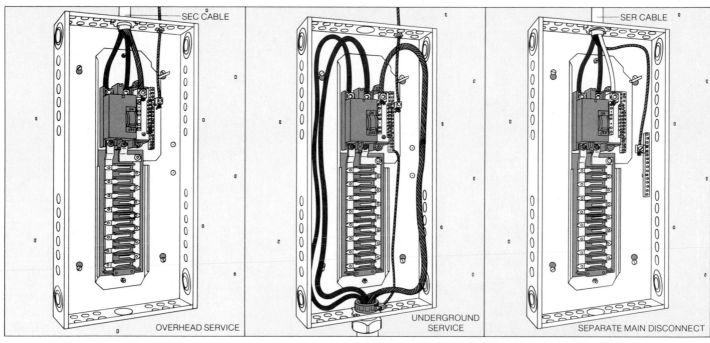

OVERHEAD SERVICE — UNDERGROUND SERVICE — SEPARATE MAIN DISCONNECT — SEC CABLE — SER CABLE

8 Wiring the ground/neutral block. Pull branch-circuit wires into the panel box as far as you can. Install lock nuts on the cable connectors and strip any exposed cable sheathing back to the connectors. Gather the bare ground wires into groups of two or three and run each group into a lug on the bottom row of the ground/neutral block, bending any slack into a long loop. To bring the grounds from the far side of the panel to the block, tape them into two bundles and route them across the top and bottom of the panel. Caution: place them so they cannot touch hot lugs or power buses. Run these grounds to the block in pairs.

Strip ½ inch of insulation from each of the white neutral wires and run them one by one into separate lugs in the empty portion of the block, bending the slack into loops. Tape neutrals from the far side of the panel into two bundles and run them alongside the grounds to the block; then separate them for individual connections.

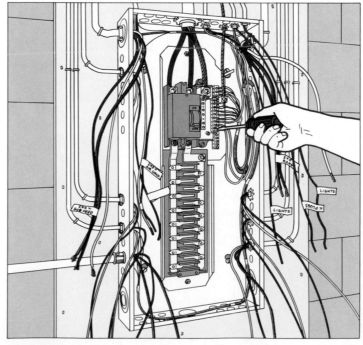

9 Wiring the breakers. Snap the circuit breakers into the panel and loop the hot branch-circuit wires to their breakers, leaving as much slack as possible for later wiring changes. Strip ½ inch of insulation from the end of each hot wire, insert the wire into its breaker lug and tighten the setscrew. Write the purpose of each circuit on the chart on the panel door and remove the tape labels from the wires. Remove knockouts from the safety cover at the breaker positions and screw the cover to the panel.

With both the main and branch circuit breakers turned off, have the power restored. Turn the main breaker on, then the branch circuit breakers, one by one. If any of these breakers trip, switch off the main breaker, cut the power to the panel, and find and correct the faulty wiring in the panel or the circuit.

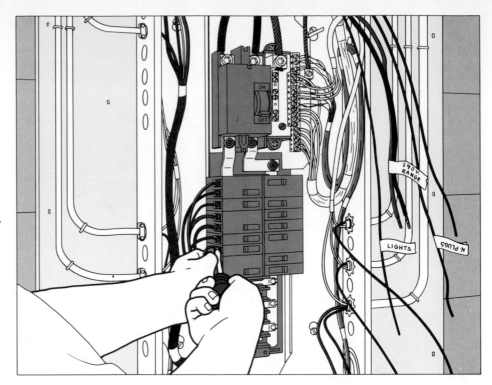

The Special Safeguard of a GFI

Every year, hundreds of people are killed by electricity that leaks to a ground in amounts too small to trip a circuit breaker or blow a fuse. Most of these people could be saved by a ground-fault interrupter (GFI), an electronic device that detects leakage by comparing the current in a hot wire with the current in the accompanying neutral wire. If the two differ by as little as .005 ampere, a magnetic switch in the GFI disconnects the circuit in ¹/₄₀ of a second, before any harm can be done.

The National Electrical Code does not require GFI protection unless you are installing new outlets in certain electrically dangerous locations—outdoors, in bathrooms and garages and at fixtures that supply swimming pools and fountains. However, this extra safeguard is a worthwhile addition to existing circuits in those locations. Either of two versions of the devices can be used, depending on needs: GFI receptacles or GFI circuit breakers.

GFI receptacles are wired into indoor or outdoor outlet boxes; they are usually installed to protect outlets supplied by existing, unprotected circuits. GFI circuit breakers, wired into a service panel or subpanel *(right),* protect entire circuits, combining the functions of a conventional circuit breaker and a GFI. A GFI breaker should be used only in a circuit limited to outlets needing ground-fault protection—the combined electrical leakages from other loads in a general-purpose circuit could cause spurious tripping. For a cable run longer than 200 feet, use a GFI receptacle; natural leakage along the cable might trip a GFI circuit breaker.

False tripping is rarely a problem in modern GFIs. If it occurs, look for faulty wiring in the circuit—bad connections at junction boxes can leak enough current to trip a GFI, as can staples that are crushing cable insulation. And test appliances for leakage just as you would for a short circuit: unplug them all from the circuit and reconnect them one by one. Old appliances with heating elements or motors—automobile block heaters, hair driers, fluorescent light ballasts, refrigerators, electric drills and the like—are likely offenders.

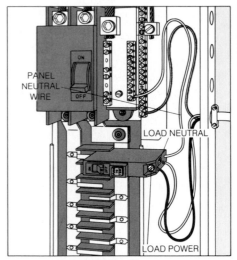

Wiring a GFI circuit breaker. With power cut off to the service panel and the GFI switched off, snap the breaker into the panel board and connect the black hot wire of the circuit to the breaker lug marked LOAD POWER, the white neutral wire to the lug marked LOAD NEUTRAL and the GFI's white, prewired "pigtail," marked PANEL NEUTRAL, to a lug on the panel's neutral block. After power is restored press the test button to simulate a ground fault. The breaker should trip immediately; reset it by simply flicking its switch to "on." Caution: the GFI is a delicate device that may deteriorate if it is not tripped regularly. Test it at least once a month and replace it if it fails to trip.

New Ways to Ground a System

Your electrical system needs a ground—a bare copper cable that carries off surges of current produced by short circuits. The cable leads from your service panel to an electrode—a metal pipe or rod buried in the ground. Caution: although in theory the grounding cable does not carry current, in reality it sometimes does. Always have power disconnected at the pole or the meter *(page 50)* before working on the grounding system.

You usually will have to put in a larger cable when you upgrade your electrical service. The National Electrical Code requires No. 8 copper for 100- and 125-ampere services, No. 6 for 150 amperes and No. 4 for 200 amperes, but No. 4 cable is generally the simplest to install whatever the size of your service, because most local codes require that smaller cables be encased in conduit.

At one time, an underground metal water pipe was the usual grounding electrode, but it is falling out of favor as utility companies replace metal water meters, pipes and couplings with nonconducting plastic ones. If your old service panel was grounded to a water pipe you are permitted to ground the new one the same way, but the Code now requires that you supplement the water pipe with a second ground electrode. This can be a gas pipe, provided (a) the gas company permits such a use, (b) your house is supplied by underground metal pipes and (c) a jumper wire carries the ground path around the gas meter *(Step 2)*; otherwise the alternate ground must be a rod driven into the ground outside your house. The rod must be at least 8 feet long (10 in Canada), and is generally made of ½-inch copper-clad steel.

If your old service panel was already grounded to a driven rod, or to a rod or copper wire extending through the concrete footing of the house foundation, you can continue to use this same ground electrode. But in such an installation the Code requires that any interior metal piping be grounded as well. This connection, called bonding, carries any current that leaks to the pipes directly to the ground; otherwise the entire piping system might be energized.

Connecting to Utility Pipes

1 Grounding to utility pipes. With the main circuit breaker switched off, clean a 2-inch section of the cold-water pipe with No. 220 sandpaper near the point where water and gas services enter your house and tighten a grounding clamp around the pipe; install a second clamp on the metal gas pipe if your gas company permits. Run the grounding cable from the service panel or main disconnect through the nearer clamp lug and on to the farther clamp, then tighten both lugs.

If your house is supplied by plastic water or gas pipes, or if the gas company does not allow grounding connections, drive an electrode *(overleaf)* to provide the secondary ground.

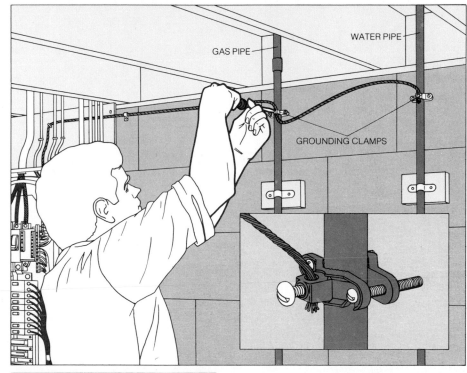

GAS PIPE
WATER PIPE
GROUNDING CLAMPS

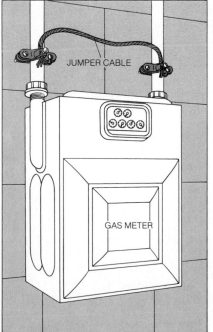

JUMPER CABLE

GAS METER

2 Jumping the meter. If your water or gas meter is located in the basement or within 10 feet of the house, install grounding clamps around the incoming and outgoing pipes and run a jumper of grounding cable between them *(left)*, leaving enough slack to allow access to the meter. Do the same at any plastic pipe fittings. No jumpers are needed for meters and fittings farther than 10 feet from the house. Label the cable with a red cardboard tag so a meter repairman will not be tempted to disconnect it.

Connecting to Buried Rods

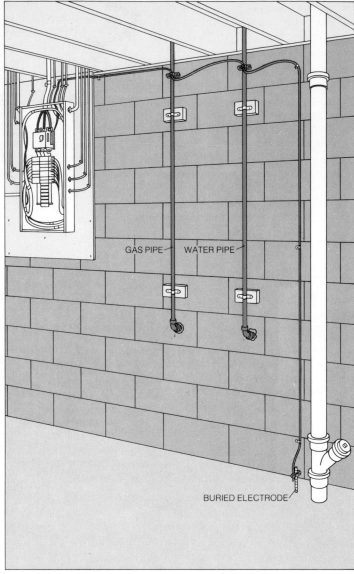

GAS PIPE   WATER PIPE

BURIED ELECTRODE

Driving a ground rod. Place the rod 18 inches from the foundation wall and drive it straight down with a small sledge hammer until only 6 inches protrude aboveground (in Canada, drive it underground). Attach the ground clamp and cable. Make a temporary barricade from the rod to the house with two 1-by-2 stakes and a horizontal 1-by-6 (*inset*) to keep people from tripping over the assembly; later you can conceal it with shrubbery. Inside the house, bond the ground cable to a metal water pipe and, where permitted, to the gas pipe (*right*).

If the rod hits the rock with more than 4 feet protruding from the ground, pull it up, attach the ground clamp and cable and bury the assembly horizontally in a trench at least 2 feet deep. If less than 4 feet protrude, bend the rod to a right angle and bury the horizontal section in a trench 18 inches deep.

A built-in electrode. If the ground cable from your old service panel ran to a length of steel reinforcing rod or solid copper wire protruding from the foundation, the new cable must also be connected to it. Run the new ground cable from the panel box through a grounding clamp placed around a metal cold-water pipe to bond the piping system to the ground; if the gas company permits, do the same at a metal gas pipe. Then run the cable to the protruding electrode, connecting it to a rod with a ground clamp or to a wire with a pressure connector called a bug (*page 61, Step 4*).

Blocking a Bolt of Lightning

When lightning hits a power line or strikes nearby, a pulse of enormous voltage surges along the line in a few millionths of a second. Devices called lightning surge arresters on the power lines usually block the surge, but if an arrester fails or if lightning strikes between your home and the nearest arrester, your television set may explode, electricity may arc out of your receptacles, and switch and breaker contacts may fuse. Even underground services are vulnerable—lightning that strikes a tree often jumps from the roots to underground cables. Such surges, rather than direct strikes, are responsible for three quarters of the insurance claims for lightning damage.

You can protect your house from surges by hooking your own arrester to the cable that supplies your service panel. The arrester shunts surges safely to ground before they can enter your electrical system. It does not, however, protect your house from direct lightning strikes—only a professionally installed lightning-rod system can do that.

An arrester works like a pressure-relief valve in a steam system. It has a connection to ground, an escape route that includes a space called a spark gap, which acts like the weight or spring in a relief valve. The resistance of the gap is sufficient to keep ordinary currents from escaping to ground; such currents flow past the arrester without hindrance. But a very high voltage, such as that produced by lightning, jumps the gap, permitting the surge current to drain away to ground.

While an arrester safeguards house wiring against surges of 2,000 volts or more, lesser voltages can harm such delicate equipment as television sets, stereo amplifiers and microwave ovens. Many of them contain protective devices called varistors (short for "variable impedance resistor"), which work like a lightning arrester but at lower voltages. You can also buy a plug-in "voltage spike protector," which contains a varistor, for any receptacles that serve electronic appliances. However, spike protectors can handle only surges less than 2,000 volts; always use them in conjunction with an arrester at the service panel.

Installing a Surge Arrester

1 **Stripping the supply wires.** Test the supply lugs to make sure that power to the panel is disconnected (*page 23*), then pare away ¾ inch of insulation from each service entrance cable hot wire. To remove the insulation without nicking the wire, score two circles ¾ inch apart around its circumference with a hooked electrician's knife, then pull away the insulation along the wire between the scored lines.

Fasten the arrester into a top knockout with a lock nut and run its white wire to a lug in the ground/neutral block, trimming the wire to make it as short and straight as possible.

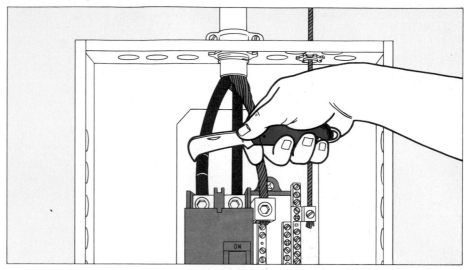

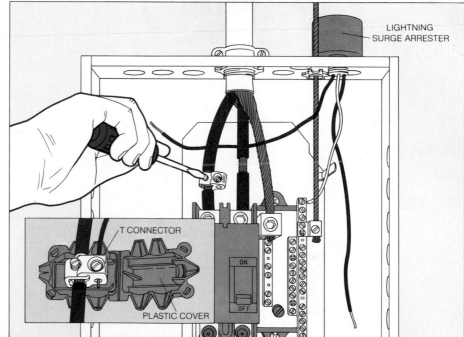

LIGHTNING SURGE ARRESTER

T CONNECTOR

PLASTIC COVER

2 **Tapping the hot wires.** Buy a T connector (*inset*) to fit the supply wires and the arrester leads, and clamp it over the bared section of one hot wire with its setscrew. Connect one black wire of the arrester to the lug on the T connector, trimming the wire to make it as short and straight as possible. Snap on the hinged, insulating plastic cover that comes with the T connector. Install another T connector on the other hot wire.

A Shutdown for Safety

A main shutdown or disconnect, with fuses or circuit breakers that can stop the flow of current throughout a house, is an important safeguard: the National Electrical Code requires that disconnecting devices be installed near the point where power enters your house. You do have a choice, though, on its precise location.

Most main disconnects are located at the top of the main service panel, above the smaller fuses or breakers that protect individual branch circuits. This location has its disadvantages. Electricians usually install the service panel in a basement, which is why homeowners must grope through dark cellars when fuses blow or breakers trip. Another disadvantage is that hot terminals stay in the panel even when the main breakers or fuses are shut down; if you must work on the panel, you should have an electrician pull your meter *(page 50)* or have the electric company turn the power off at the pole.

In a more convenient arrangement, the main disconnect is set in a small separate panel of its own, at the point where the service entrance cable enters the house; a cable links the disconnect to a subpanel containing the breakers or fuses of the branch circuits. You can cut all power at the disconnect and work safely on branch circuits in the subpanel. The subpanel can be placed wherever you choose, for the connecting cable from the main disconnect can run to any room and if you change the location of your meter, it will be easier to install a main

disconnect box at the new location than to move the entire service panel.

Small main disconnect panels are available at electrical-supply stores as conventional boxes for indoor installation or as weatherproof boxes for the outside installation shown on these pages. Be sure that the box you buy can be padlocked through flanges on the bottom so that no one can turn on the power while you are working at the subpanel.

The main disconnect is connected to the meter box by a conduit containing two insulated hot wires and a bare neutral wire—the same three wires that run through an SEC cable. Since these wires are enclosed by the conduit, you can safely strip the cable sheathing from them and push them through the conduit individually. Power flows from the main disconnect to the branch-circuit panel through conduit containing the service entrance round (SER) cable—two hot wires, an insulated gray neutral wire, and a bare wire that grounds the service panel to that of the main disconnect. As the primary power panel, the disconnect must be grounded with the rest of your electrical system *(Step 4, opposite)*.

Have an electrician pull the meter or the electric company turn the power off at the pole before you install the main disconnect. If you install a main disconnect and a new meter at the same time, have the service disconnected, install the new meter and the disconnect, then restore temporary service *(pages 51-53)*.

1 **Installing the box.** Mount the main disconnect box on the wall, locating it 12 inches from the meter box and aligning its side knockout with the one on the side of the meter box. Connect the two knockouts with a 12-inch length of 2-inch plastic conduit, adjusted to fit with threaded terminal adapters and held inside the boxes with lock nuts and bushings.

Remove the sheathing from an SEC cable and run the separate wires through the conduit; allow for 16 inches of extra wire in each box. Use 4/0 aluminum or 2/0 copper wire for a 200-amp service, 2/0 aluminum or No. 1 wire for 150-amp service and No. 2 aluminum or No. 4 copper wire for a 100-amp service. Connect the wires in the meter socket *(page 53, Step 1)*.

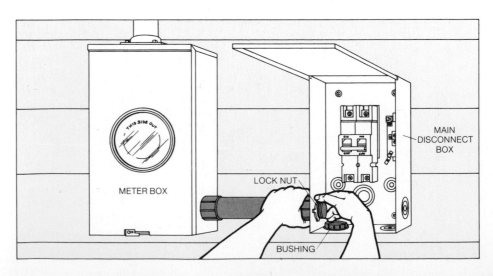

2 **Running the cable.** Make a 2-inch hole through the wall of the basement (page 57) and fit the hole with a 90° elbow section of 2-inch conduit, threaded on both ends; cut a length of 2-inch conduit to fit between the top of the L and the knockout on the bottom of the main disconnect box. Screw the threaded end of this conduit into the elbow and fasten the cut end to the knockout with a threaded terminal adapter, a lock nut and a plastic bushing.

Push SER cable from the main disconnect box through the 2-inch conduit to the service panel inside the house, leaving 16 inches of cable in the main disconnect box. Strip 14 inches of sheathing from the cable (page 15), exposing the insulated gray neutral wire, the two black hot wires and the bare wire.

RIGID CONDUIT

ELBOW

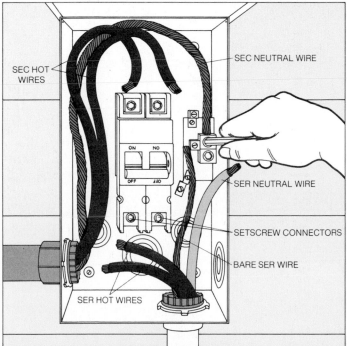

SEC HOT WIRES

SEC NEUTRAL WIRE

SER NEUTRAL WIRE

SETSCREW CONNECTORS

BARE SER WIRE

SER HOT WIRES

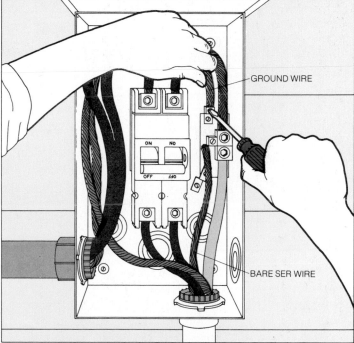

GROUND WIRE

BARE SER WIRE

3 **Wiring the disconnect.** Cut the SEC and SER wires to reach their setscrew connector without excessive slack; strip an inch of insulation from the black and gray wires and twist the bare wires into continuous strands (page 15). Fasten the bare SEC wire to the connector at the right of the disconnect panel and the gray wire to the connector below; use a hex wrench to tighten the screws that hold the wires. Fasten the black SEC wires to the connectors at the top center of the panel and the black SER wires to the connectors at the bottom.

4 **Grounding the system.** Run a ground wire from the grounding electrode (pages 45-46) to the main disconnect, and fasten it to the connector at the top right of the disconnect panel. Disconnect the ground wires from the service panel and remove the bonding screw. Install a new grounding bar (page 43, Step 7) in the panel and run the ground wires to it. Finally, attach the bare SER wire to the connector directly below the one you used for the ground wire in the main disconnect; attach the other end of this wire to the new grounding bar in the service panel.

How a Meter Is Unplugged

To make sure that all the power coming into a house is measured, electric companies make the meter an essential part of the electrical path from the power lines to the service panel. Removing the meter to break this path is the easiest way to turn off the power in the whole house—an essential safeguard for attaching branch circuits at the service panel.

The meter is mounted in a chassis that resembles a large plug; the prongs of the plug fit into a meter socket permanently mounted on a house wall. When electricians remove this plug they say they are "pulling the meter." Homeowners who are experienced in working with electricity sometimes pull their own meters, but the job is more often left to a professional—the power that flows through the meter is unfused and extremely dangerous. If you decide to pull your own meter, obtain an electrical permit and notify your power company beforehand. Never attempt the job yourself if the wires to or from the meter are damaged.

Before pulling a meter, an electrician first turns off the main circuit breakers or pulls the main fuses in the service panel. Then, standing on a sheet of ¾-inch plywood covered with a rubber mat to insulate him from the ground, and wearing thick rubber gloves certified by the Underwriters' Laboratories (UL), he cuts a twisted wire seal that locks a collar clamp around the meter. Next, he removes the clamp screw with a plastic-handled screwdriver, opens the collar and pulls it away from the meter *(top)*.

To remove the meter, the electrician grasps it on both sides and pulls it down and out to release it from the clips in the socket *(center)*. As soon as the meter is free, he immediately inserts a glass safety plate into the old meter collar, sets the assembly over the exposed meter socket *(bottom)* and screws the collar clamp back into place. The glass plate is essential to keep anyone from touching the hot terminals in the socket.

Finally, he checks the service panel with a voltage tester to be sure that the power has been cut off inside the house. If it has not, the electric company must check the panel and meter before any further work is done.

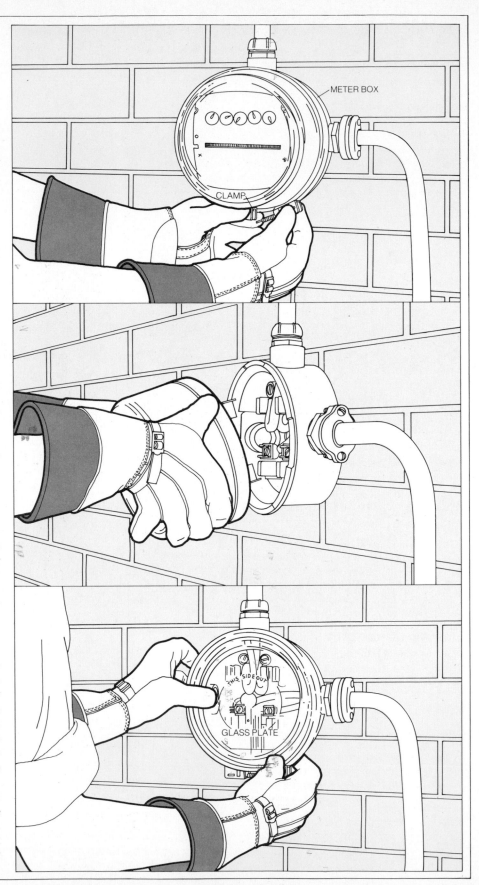

Outside: Replacing the Utility Company's Meter

If you upgrade a panel from 60 amperes to 100, 200 or more *(pages 40-44)*, you must also change the meter to handle the increased power load. These jobs should be done at the same time—the meter pulled, the panel upgraded and the new meter and meter socket installed.

The power company will shut down the power while you work on the new service panel, but in many localities it may be as much as a week before power can be restored. An alternative is to have an electrician pull the meter before you begin work. Then you can change the panel and detour the power through the old meter into a new meter socket and the new service panel. After the final inspection has been completed, the power company will take out the old meter and make a permanent hookup.

Before you begin this job, you must get a permit and read your local codes regarding meter and service changes. When you have the permit, go to the electric company and see what services they offer. Normally, they will supply you with a meter and meter socket designed to handle the increased amperage, a glass plate to cover the hot terminals in the old meter box temporarily, and weatherproof sockets for the service cable.

You will find the other items you will need at an electrical-supply store—two 6-inch lengths of No. 4 copper wire to serve as temporary jumpers in the new socket, new cable, metal clamps to pin the cable to the wall of your house and—if you are using aluminum cable—anti-corrosive paste. You may also want to get certain hand tools to simplify the job, since you will be working without electricity—for example, a twist drill that can pound a carbide bit into any masonry surface and a drive-pin set to clamp the cable to the wall. If the cables to and from the meter are exposed to physical damage—near a driveway, for example—the code requires the protection of conduit *(pages 54-57)*.

If the electric company does not supply the meter, it is usually because the local building authorities require that you install a main disconnect, like the one shown on pages 48-49. If this is the case, you may wish to purchase a special prewired box that contains both the meter and the main disconnect. This combination box allows you to shut off the power to the panel and branch circuits without pulling the meter.

Many power companies have introduced meter sockets with compression-type connectors, which can be wired only with expensive hydraulically operated pinchers. If you have such a meter, you will need special lug connectors like the one shown on page 53, Step 1, to make a sound and secure connection until the power company arrives. Caution: never wrap the wire around a connector—it may come loose and energize the housing of the meter.

While your temporary service is in operation, do not use more amperage than your old service panel could handle—you may overload the old supply lines from the electric company's transformer. After the work is done and inspected by the local building authorities, the electric company will check its lines to be sure they are adequate for the increased amperage and make the service permanent by taking out the old meter and running the cable directly to the new one.

Mounting a Meter Socket

1 **Mounting the new socket.** Use a twist drill to make mounting holes in a brick or cinder-block wall. After making sure all power to the house is off by testing the hot and neutral leads in the service panel, pull the old service cable through the wall. Then set the new meter box against the wall, 12 to 18 inches away from the old meter and 3 to 6 feet above the ground. When the box is level and plumb, mark the location of its screw holes on the wall. Set the bit of the drill against a mark and hammer the handle until the bit is engaged. Twist the drill 60° clockwise and strike again. Twist and strike until you have a hole 2 inches deep, then repeat the procedure at the other marks. Fasten the box with lead anchors and 1½-inch wood screws. In a wooden wall, simply fasten the meter box with 2-inch wood screws.

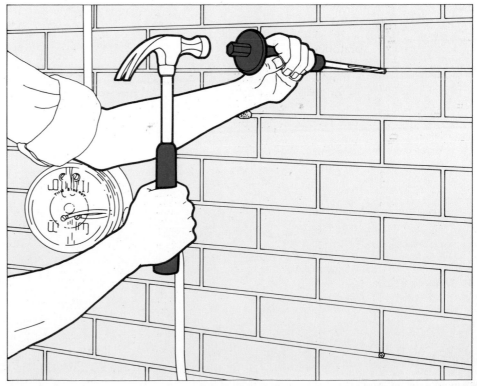

2 **Making a service drip.** Pull about 8 feet of service cable from the new service panel out of the house through the hole left by the old cable and bend the new cable down toward the ground, then up to the new meter box, in a curve that lies flat against the wall. Seal the hole in the wall with premixed sealing compound, available at electrical-supply stores.

3 **Weatherproofing the wires.** Slide the weatherproof socket 18 inches up the free end of the new service cable, tighten the socket firmly around the cable, then push the cable through the hole in the side of the new meter box until the socket meets the box. Slide a lock nut over the end of the cable and tighten it around the threads of the weatherproof socket. Push about

18 inches of the old service entrance concentric (SEC) cable wires through the hole in the top of the box and pack the hole with sealing compound. Strip 16 inches of sheathing from the free end of the cable with a hooked electrician's knife, exposing the two black hot wires and the unsheathed neutral wires. Gather the neutral wires and twist them together *(page 15, bottom)*.

NEW METER BOX

OLD METER BOX

OLD SEC CABLE

NEW SEC CABLE

WEATHERPROOF SOCKET

LOCK NUT

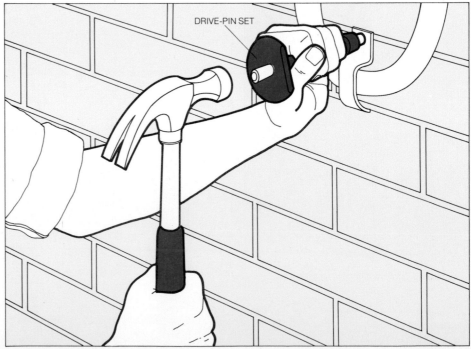

DRIVE-PIN SET

4 **Securing the cable.** Set a metal clamp over the curve in the new cable, position a drive-pin set fitted with a 1-inch pin over the hole in the clamp, and strike the head of the pin set sharply with a heavy hammer to drive the pin into a mortar joint of a brick or cinder-block wall. Install as many clamps as you need to hold the cable firmly against the wall.

On a wooden wall, fasten the clamps with round-headed wood screws 1 inch long.

Wiring the Socket

1 **Wiring the socket.** Run the wires of the new cable to the setscrew connectors at the bottom of the socket with the neutral wire in the center; if your socket has bolts rather than setscrew connectors, install special lugs (*inset*) over the bolts to secure the wires. Hook the old cable wires to the top connectors, neutral in the center.

2 **Jumping the new meter.** Turn the ends of a 6-inch length of No. 4 copper wire at right angles to make a jumper, and insert the ends of the jumper into the clips at one side of the new meter socket. Plug a second jumper into the clips on the opposite side of the socket.

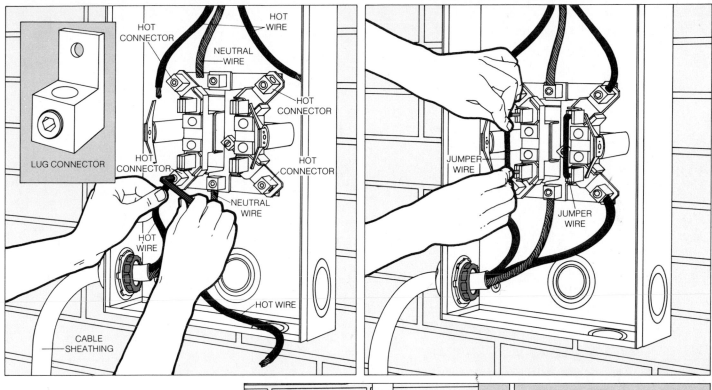

LUG CONNECTOR

HOT CONNECTOR
HOT WIRE
NEUTRAL WIRE
HOT CONNECTOR
HOT CONNECTOR
HOT CONNECTOR
NEUTRAL WIRE
HOT WIRE
CABLE SHEATHING
HOT WIRE

JUMPER WIRE
JUMPER WIRE

3 **Restoring temporary service.** Call an electrician to reconnect the old meter. Wearing rubber gloves, he will remove the collar-and-glass assembly from the old meter, fasten the glass to the cover of the new meter with tape and cover the new meter box (*right*). Then he will plug the old meter back into its socket, pushing the bottom into place first, and reseal the collar. Caution: the old meter should never be replaced until the new meter is securely covered and the protective glass plate installed (*far right*).

Check to be sure that power has been restored to the service panel inside the house, then arrange for an electrical inspection of the entire service and meter installation.

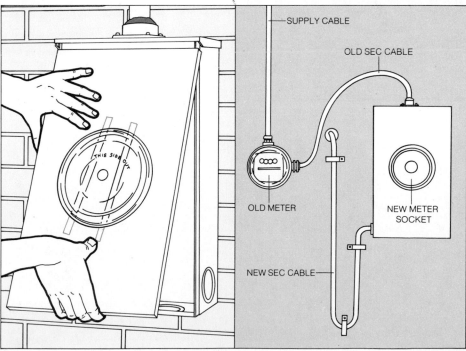

THIS SIDE OUT

SUPPLY CABLE
OLD SEC CABLE
OLD METER
NEW METER SOCKET
NEW SEC CABLE

Protecting Conductors with Conduit

Whenever you route power outside your home, you will need more than the skills of an electrician. The National Electrical Code requires that wire or cable must be encased in hard pipes called conduit wherever there is a serious risk of damage—in a passage through a house wall or roof, in a run that emerges from the ground to connect to a meter, in shallow underground runs and in any underground run that passes beneath a driveway, concrete slab or sidewalk. The job of cutting and fitting these pipes resembles nothing so much as plumber's work.

Start the job by choosing the type of conduit—heavy wall, thin wall or plastic—that is best suited to your needs. Rigid, or heavy-wall, conduit of galvanized steel, the most common type, has a strength and a resistance to corrosion that make it particularly suitable for runs subject to mechanical abuse, the weight of passing vehicles or bad weather. It comes in 10-foot lengths threaded at each end and provided with a coupling that links two sections together. Shorter lengths, called nipples, are used for connections through walls and for short runs.

Electrical Mechanical Tubing (EMT), or thin-wall conduit, lighter and easier to work with, is useful for relatively protected locations, such as a connection from a

Fittings for Turns and Connections

Bends. A threaded condulet routes conduit in curves above the ground and has a small hatch that can be opened to adjust cable or wires as they are pushed around bends. The Type LL condulet shown here opens on the left side; condulets with openings on the back (Type LB) and on the right side (Type LR) are also available. A threaded elbow, larger but less expensive than a condulet, comes in 90° and 45° bends. Unlike a condulet, an elbow can be buried below the ground; it is especially useful for running cable up out of an underground service.

A threaded offset adapter can be used to route conduit to a panel box that is set out a slight distance away from the house wall. If you must run conduit around a plumbing pipe, gas line or other obstruction, use two adapters joined by a length of rigid conduit.

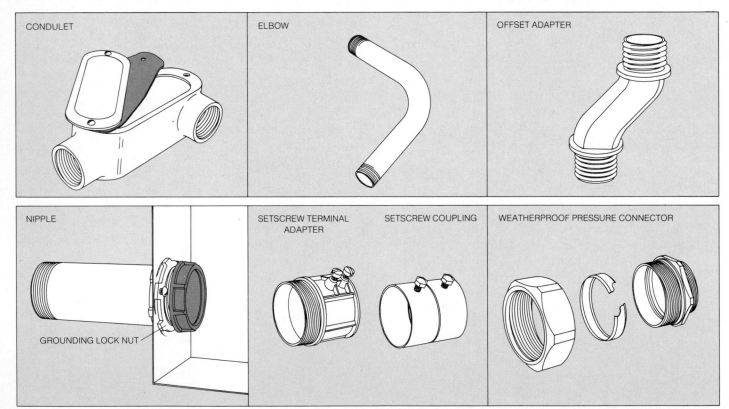

CONDULET

ELBOW

OFFSET ADAPTER

NIPPLE

GROUNDING LOCK NUT

SETSCREW TERMINAL ADAPTER

SETSCREW COUPLING

WEATHERPROOF PRESSURE CONNECTOR

Connectors. A galvanized-steel nipple, threaded at both ends and available in 4- to 36-inch lengths, directs cable through walls or small spaces. At a knockout in a service panel or meter box, the nipple is secured by a lock nut screwed against the outside of the box and a grounding lock nut screwed so tightly against the inside of the box that its tangs bite into the paint; a setscrew holds the inside lock nut permanently in place and a plastic bushing screwed over the exposed conduit threads protects the cable from abrasion. A threaded setscrew terminal adapter connects EMT conduit or a cut end of rigid conduit to panel boxes or condulets. Tightening the setscrews fastens the adapter to the conduit; the connection, like that of a nipple, is made with lock nuts. Join two lengths of EMT conduit with an unthreaded setscrew coupling, which fits like a sleeve over both conduit ends. A weatherproof pressure connector joins unthreaded pieces of conduit in damp locations. Tightening the large nut squeezes a metal flange around the conduit to form a waterproof seal; a lock nut secures the conduit to a box.

meter to an outside panel or from an underground service to a meter. It comes in unthreaded 10-foot lengths linked by setscrew or pressure connectors.

Lightweight plastic conduit, the least expensive and the easiest to use, is sold in two varieties. Polyvinyl chloride is best used aboveground; in areas exposed to direct sunlight, cover the conduit with two coats of latex paint to prevent deterioration. High-density polyethylene is suitable for underground runs that will not be exposed to great stress. Both types

come in unthreaded, 10-foot lengths that can be joined with glue rather than with threaded or setscrew connectors.

The conduit you use, usually 1 or 2 inches in diameter, will be difficult or impossible to bend, but a great variety of angled connectors and special fittings makes bending unnecessary. Always begin an installation by mapping the route that the conduit will follow and choosing the straight lengths and fittings you will need. If you have many lengths of rigid conduit to cut you can save time by using

a pipe cutter, available at tool-rental agencies. The same agencies can provide small jackhammers called hammer drills for cutting holes in concrete.

To run conduit through a hole in an exterior wall, push the conduit through the wall and the cable through the conduit—a mixture of powdered detergent and water smeared on the cable will make it slide easily. Seal the hole around the conduit with premixed patching concrete, and waterproof each end of the conduit with sealing putty.

Three Basic Conduit Routes

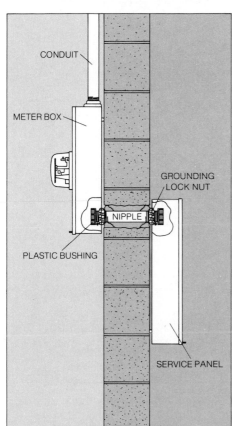

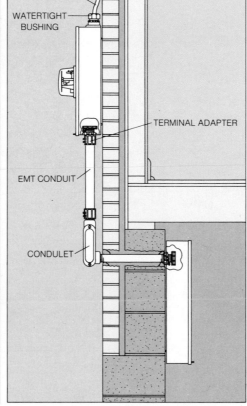

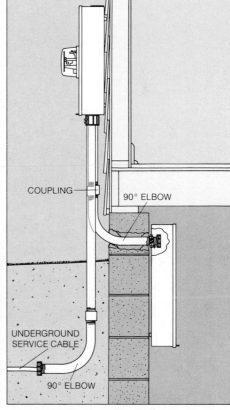

A single nipple. When a service panel is mounted inside the house directly behind the meter outside, join the two boxes by running cable through the house wall in a rigid conduit nipple threaded on both ends. Connect the nipple to the outside of each box with a lock nut; screw a grounding lock nut and a plastic bushing to the nipple ends inside each box. The lock nut completes the grounding circuit between the boxes and the conduit; the bushing protects the cable from abrasion. Seal the hole around the nipple with premixed patching concrete.

Down from above. Cable from an overhead service runs down the house wall through a watertight bushing into the meter. A second cable runs out of the meter through EMT conduit into a condulet; the unthreaded EMT is connected to the condulet and meter with setscrew terminal adapters. The cable runs through the house wall to the service panel in a rigid conduit nipple, which screws into the condulet and the panel box. The hole in the wall is sealed with patching cement and the end of the conduit waterproofed with sealing putty. A lock nut secures the conduit outside the service panel; a grounding lock nut secures it inside.

Up from below. An underground service entrance cable turns upward in a 90° galvanized-steel elbow, connected to a 4-foot length of rigid conduit that runs up the house wall into the meter box. At the lower end of the elbow, sealing compound waterproofs the cable and a threaded plastic bushing protects it from abrasion. The conduit, cut from a 10-foot length, has no threads on the upper end; a setscrew terminal adapter connects it to the meter box and inside the box it is grounded with a grounding lock nut. In a similar conduit-elbow arrangement, a second cable leaves the meter, enters the house above grade and runs into the service panel.

Cutting Conduit to Fit

1 **Cutting through the conduit.** Secure the conduit in a vise and tighten the jaws of a pipe cutter around it until the cutting wheel presses on the conduit (*below, left*). Swing the cutter completely around the conduit, then tighten the jaws again to dig the wheel more deeply into the metal. Repeat until the conduit breaks in two.

To cut through conduit with a hacksaw (*below, right*), wrap a length of masking tape around the conduit to prevent the saw from slipping—overlap the ends of the tape—and mark the tape at the cutting point. Begin sawing with the hacksaw blade parallel to the floor; then, after a few strokes, angle the blade toward the floor. If the blade sticks, press down on the projecting end of the conduit with your free hand to open the cut a little. Use a blade with 18 teeth per inch for rigid conduit, 32 teeth per inch for EMT and plastic conduit.

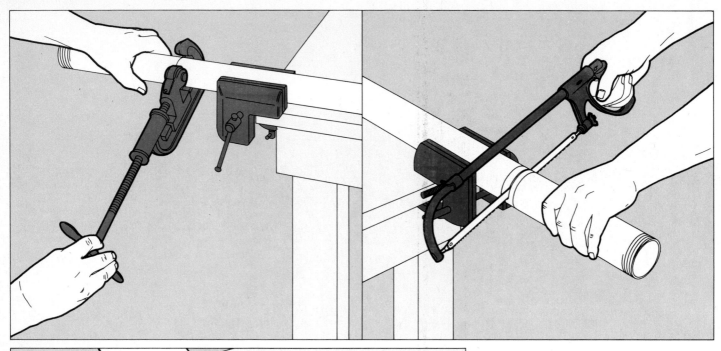

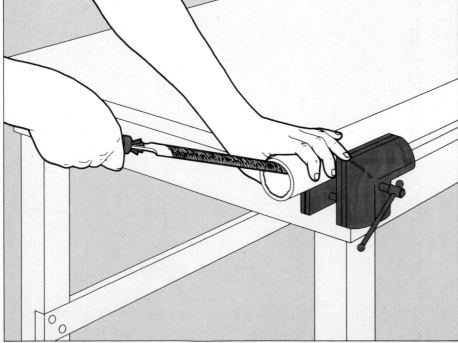

2 **Removing the burrs.** Smooth the burrs on the cut end of rigid conduit or EMT with a half-round file. On plastic conduit cut the burrs off with a sharp knife, then bevel the outside edges of the conduit to fit it into a coupling for gluing.

A Passageway Through a House Wall

In cinder block. Trace the end of a piece of conduit set about a third of the way across a cinder block over one of the hollow centers, tape the edges of the circle and drill a series of ¼-inch holes just inside the circle. Break out the spaces between the holes with a cold chisel.

Center an extension bit in the hole you have made and drill through the inside of the house wall. Inside the house, using the small hole as a center, scribe a circle the size of the conduit with a compass, then follow the techniques described above to complete the conduit hole. Push the conduit through the wall and seal it in place with premixed patching cement.

In wood. Locate the studs inside the wall and drill a hole in the outer wall between two studs, using a hole saw as wide as the conduit. Remove any insulation in the path of the conduit hole; then center an extension bit in the hole, drill a pilot hole through the interior wall and complete the conduit hole inside the house.

Push a nipple through the wall—let the threads protrude on both sides—and tighten lock nuts at the ends. Waterproof the inside and outside of the wall with sealing compound.

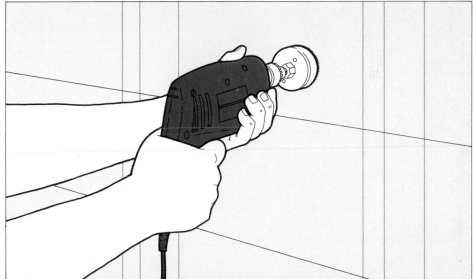

In concrete. Start the hole with a punch and deepen it with a heavy-duty hammer drill fitted with a carbide-tipped hole saw as wide as the conduit. Apply heavy pressure until the hole saw jams in the concrete, then chisel out the concrete at the center of the hole. Repeat as necessary to drill completely through the wall.

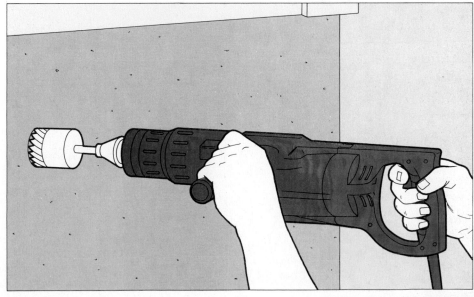

The Final Step: Bringing Power to the Meter

To bring power to your house, an electric company runs a group of wires called a service drop from its overhead supply lines to a point where a splice can be made to your service entrance concentric (SEC) cable. Older three-wire service drops consist of three separate wires; newer ones, called triplex, have a bare neutral wire with two insulated hot wires wrapped around it. In both installations, the wires are dangerous; the National Electrical Code requires that their connections must be high above the ground, well out of the way of cars and people.

The connections themselves, high on the house wall or even above the roof, are made through a two-part overhead service: a set of insulators that support the service drop, and a metal hood, called a weatherhead, that protects the SEC cable from moisture. If you move your meter, you may have to move both the insulators and the weatherhead. And if you upgrade your service panel (pages 40-44), you will have to replace the meter, SEC cable and weatherhead with equipment large enough to handle the increased amperage.

The electric company may do part of this work for you: in many localities the company will attend to the overhead equipment when you have installed a new meter socket. If it will not, ask one of its representatives about the location of the new service: electrical codes set restrictions on the distance a cable can run from a utility pole to a house, and if your new service location is too far from the pole you may have to run the cable underground (pages 62-63). If you have three-wire service you may want the company to replace it with triplex. And if you are replacing your meter, you may want the company to disconnect the service at the pole. Caution: never work near a live service drop while you set up an overhead service—have the electric company cut off power.

At the house, the location of your new weatherhead and insulators must conform to the rules of the National Electrical Code. The weatherhead and the insulators must be at least 3 feet away from windows, porches and fire escapes. According to most local codes, any wires running across your property must be at least 10 feet above lawns and patios, 12 feet above driveways and 18 feet over streets and alleys. If you must raise a mast above the roof (pages 60-61) to reach the height required by code, the supply wires must run at least 18 inches above the roof and no more than 4 feet across it.

Many electrical-supply stores carry complete kits for setting up an overhead service. If you must buy the parts separately, get enough SEC cable to run from the meter to the weatherhead plus an additional 5 feet for connections, clamps to secure the cable to the wall and, of course, a new weatherhead and one or more glass insulators. For a mast, you will need a conduit (normally 2 inches for a 150-ampere service or more; 1¼ inches for 100 amperes or less; and use 2-inch conduit for any drop longer than 60 feet), circular metal clamps to hold the conduit to the house wall, a galvanized-metal roof jack to weatherproof the hole for the mast, and a steel guy wire to support a mast more than 3 feet high.

The techniques shown on the following pages may also be used to extend a branch circuit from your house to an outbuilding. For a circuit of less than 100 amperes, make the mast from 1-inch diameter conduit, bracing it with circular clamps and guy wire. Use cable rated for the amperage of the new circuit, and attach the cable to the insulators on the outbuilding with wedge clamps, available at electrical-supply houses. Form drip loops (below) at both weatherheads.

Two paths for power. Electricity comes to the house at right through a service drop consisting of three separate wires secured to glass insulators on the wall of the house; the insulators themselves are mounted on a rack attached to a wall stud. From this point the wires of an SEC cable carry power down the wall to a meter. Connections between the drop and the cable are made with wire clamps called "bugs." The cable wires run to a waterproof metal weatherhead through a drip loop that keeps water from flowing along the wires to the weatherhead. The sheathed cable between the weatherhead and the meter is secured by U-shaped clamps.

In the house at far right, the drop (in this example, a compact triplex) and the weatherhead are mounted on a service mast, a 2-inch conduit extending through the roof. A guy wire tightened by a turnbuckle helps to support the mast; a collar called a roof jack prevents water from running down the conduit to the meter. Below the roof, circular clamps hold the conduit an inch away from the house to align it with the entrance hole on the top of the meter box.

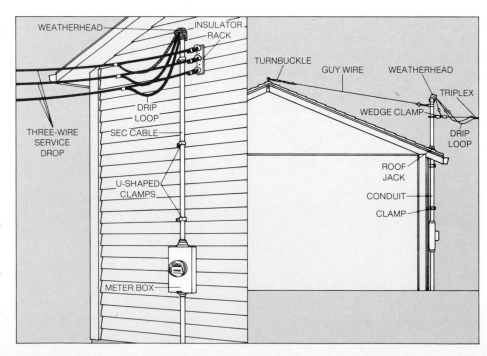

A Wall-mounted Power Line

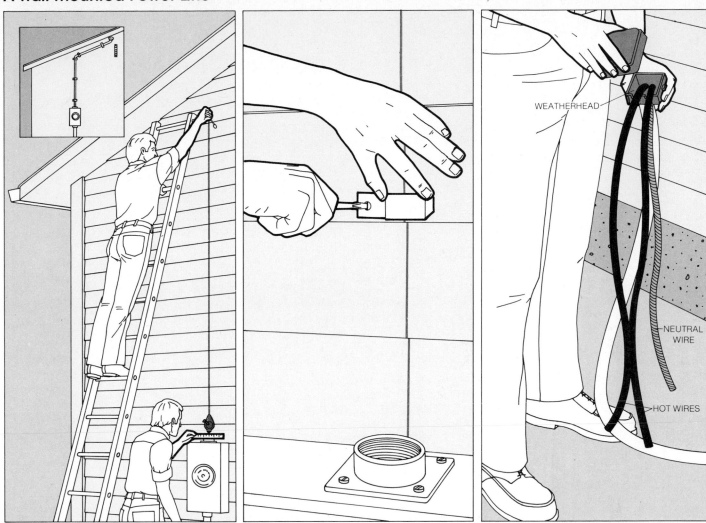

WEATHERHEAD

NEUTRAL WIRE

HOT WIRES

1 Aligning the cable. Drop a plumb line from the new weatherhead location to the center of the knockout in the top of the meter box— have a helper set a ruler across the knockout to center the plumb bob—then snap a chalk line from the weatherhead position to the knockout. To find the amount of SEC cable you will need, add 5 feet to the length of the chalk line.

If the wall directly above the meter is less than the height required by code, you can run cable as high as possible above the meter, bend it at that point, then run it higher parallel to the roof line (inset). When you install the cable, anchor the bend with U-shaped clamps.

2 Mounting the cable clamps. Center a U-shaped clamp over the chalk line 12 inches above the meter box and fasten the clamp to the wall with a 1½-inch screw. Drive the screw all the way in, then unscrew it a few turns— the clamp should be loose enough so that you can slide a cable under it. Mount additional clamps every 4½ feet along the chalk line.

In a brick or cinder-block wall, mount the cable, using a drive-pin set (page 52, Step 4) or lead shields to secure the clamps to the wall.

At the top of the chalk line, screw the support bracket of the weatherhead to the wall.

3 Wiring the weatherhead. Strip 3 feet of sheathing from one end of the SEC cable; twist the bare neutral wires into a single strand and bend them away from the insulated wires. Remove the top of the weatherhead, thread the insulated wires through the holes in the base and pull these wires out until the neutral presses against the bottom of the weatherhead. Bend the neutral against the unstripped cable and replace the top of the weatherhead.

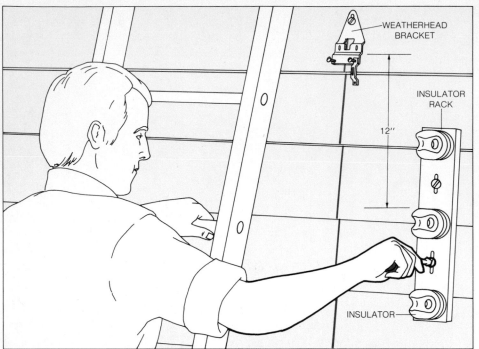

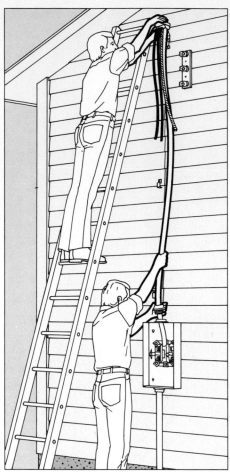

4 **Setting up the insulators.** Find the stud nearest the weatherhead bracket and, at a point 12 inches below the level of the bracket, fasten one or more insulators to this stud. Use a single insulator and screws for a triplex service drop or a rack with three insulators and lag bolts for a three-wire service drop.

5 **Installing the weatherhead.** Slide the weatherhead into the slots of its bracket and have a helper feed cable down through a waterproof connector into the top of the meter box. When the helper pulls the cable tight, slide the wall clamps over it and tighten their screws. Connect the cable at the meter socket; the electric company will connect the other end of the cable to the service drop.

A Roof-mounted Power Line

1 **Making a hole in the roof.** Drop a plumb line from the eave or fascia board to the center of the meter box, mark the eave or fascia at the top of the line and pry loose the shingle directly above the mark. Drill a hole up from the mark and through the eave. From the roof, scribe a circle the size of the conduit around the hole, then cut out the circle with a saber saw.

2 **Weatherproofing the hole.** Slide a roof jack under the loosened shingle, until the hole in the jack lies directly over the one in the eave. Seal the jack around the hole with roofing cement, then nail both the jack and the shingle to the roof with six 1½-inch roofing nails; cover the nail-heads with roofing cement. Insert the stripped end of the SEC cable into the weatherhead.

3 **Raising the mast.** Have a helper below the roof push conduit up through the roof jack to the mast height required by the code. On the roof, slide the SEC cable down into the conduit until the weatherhead rests upon the mast. Have your helper screw the conduit into the meter box, then fasten the weatherhead to the mast with the setscrews located at the bottom of the head. Attach the conduit to the wall every 4½ feet with offset circular clamps and toggle bolts. Screw an insulator clamp to the mast, 12 inches below the weatherhead, and mount the ceramic insulator for the service drop.

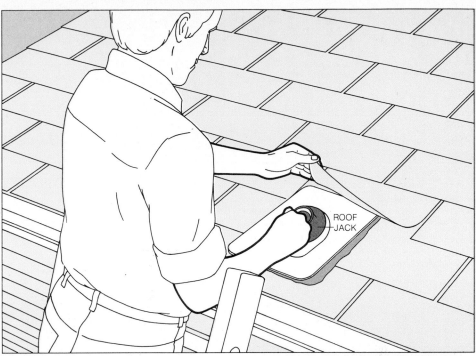

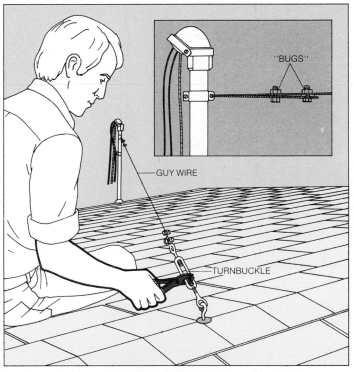

4 **Securing the mast.** Attach a circular clamp to the mast just below the weatherhead and run No. 6 steel or copper guy wire through two bug connectors and the clamp (inset) to the roof ridge. Drill a ¼-inch hole in the ridge and screw in a ⅜-inch eyebolt with a release lever. Slip a turnbuckle into the eyebolt, open the turnbuckle by unscrewing the threaded ends and fasten the guy wire to the loose end of the turnbuckle with a second set of bugs. Tighten the wire by turning the turnbuckle with a pair of pliers.

An Underground Power Line

Invisible and sheltered from the elements, underground wiring is an attractive alternative to an overhead service. What is more, the job of running wires underground has never been simpler than it is today. Recently developed Underground Service Entrance (USE) and Underground Feeder (UF) cables, both sheathed in plastic, are sturdy enough to be laid directly in a trench, without conduit. And although a short trench still must be dug with a shovel, tool-rental companies can supply you with a powered trenching machine for a long one.

You can use trenches to conceal either a main service connection or a branch-circuit cable that runs from your house to an outbuilding. In a so-called service-lateral installation for a branch circuit, the cable never appears above the ground. It leaves the house and enters the outbuilding below grade. If you replace a main overhead service with an underground cable, the part of the cable that goes aboveground to your meter should be encased in conduit.

If you plan to convert a main overhead service to an underground one, discuss your plan with your electric company; many companies will pay for the trenches, the cable or both. Even if your company does not, it will help you map a route for your trench, and will disconnect your service before you begin work and make the final connection at the pole.

Before you make a final map and start digging, consult the gas and water companies—you must stake out their underground lines on your property so that you will not cut them; in addition, stake out your septic tank, if you have one.

The National Electrical Code specifies many details of an underground installation. USE and UF cables must be 24 inches below grade for a service or circuit that is more than 30 amperes, 12 inches below grade for one that is less (Canadian code differs slightly). Under driveways, patios or sidewalks the cable must be encased in galvanized-steel conduit. In extremely rocky soil the cable must lie on a 3-inch bed of sand and be covered by 3 or more inches of sand. The protection of the sand is worth having even when it is not required; if you want it, you must dig a trench 3 inches below the code specification to put the cable at the right depth. Finally, check your local building code; some codes call for a support of drain tile, wood or concrete in addition to sand.

Choose a cable size according to the instructions on pages 124-125. When the cable is in the trench and the outlet boxes are connected, check the circuits (pages 36-37) to be sure that you have hooked up the wires correctly. Only then is it safe to bury the cable underground.

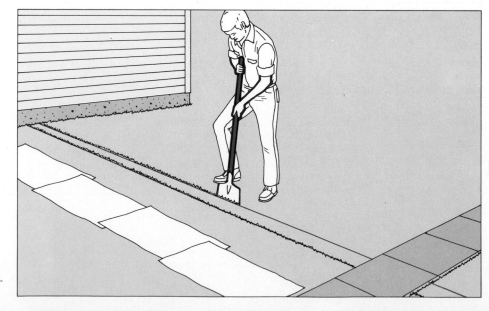

1 **Removing the sod.** Mark the path of the trench on the ground by snapping two parallel white chalk lines 8 inches apart, then loosen the sod within the lines with a flat-bladed shovel. Divide the loosened sod into 2-foot sections, lay a sheet of plastic film along one side of the path and lift the sod onto the plastic with the shovel.

At the edge of a house wall, driveway, sidewalk or patio, broaden the trench so that you can comfortably drive conduit or run a service lateral.

2 Digging the trench. Set a trenching machine at one end of the trench path with its digging teeth poised above the ground, adjust the drive speed for your soil according to the manufacturer's instructions and start the motor. Operating procedures vary; on the widely used model shown here, you must release a clutch next to the handle to start the teeth rotating, then turn a wheel on the body to lower them to the trench depth you have chosen.

When you engage the drive release, the machine will move backward. Guide it by the handles at the rear. If there is a large rock along the trench path, stop the machine, move it beyond the rock and continue trenching. Remove the rock later or, if it is too difficult to move, reroute the trench around it.

Lay a bed of sand 3 inches deep along the bottom of the trench.

3 Running conduit and cable. With a sledge hammer, drive a length of galvanized-steel conduit under an obstruction such as a sidewalk or driveway. Use 1-inch conduit for a circuit under 100 amperes; 2-inch conduit for a larger circuit. Cut the conduit a foot longer than the obstruction and cover the leading end with a closed plastic bushing; when the conduit protrudes 6 inches beyond the obstruction, file the burrs off the cut end, remove the bushing from the leading end and cover each end with a setscrew connector and an open bushing.

Lay cable in the trench and push it through the conduit; in the trench, bend the cable at 18-inch intervals to form a zigzag pattern.

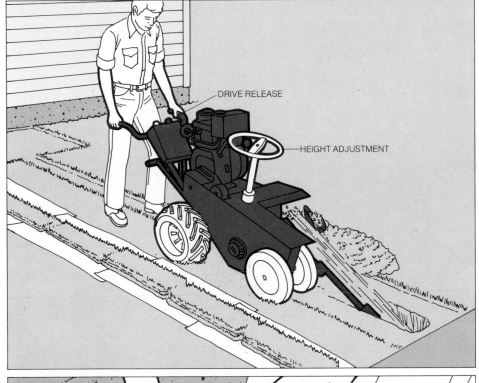

DRIVE RELEASE

HEIGHT ADJUSTMENT

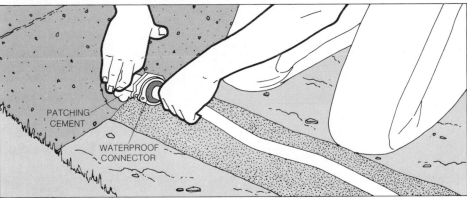

PATCHING CEMENT

WATERPROOF CONNECTOR

4 Completing the service. Make a hole through the house wall at the bottom of the trench, insert a short length of conduit threaded at both ends and glue the conduit in place with premixed patching cement. Install waterproof connectors and plastic bushings at the ends of the conduit inside and outside the house, push the cable through and caulk the conduit at both ends with a putty-like sealing compound available at electrical-supply stores. To complete the connections for a branch circuit, run the cable into an

outbuilding by the same method, have an electrician pull your meter to cut off all power to the main panel of your system, then connect the cable to the main panel in the house and the subpanel in the outbuilding.

Cover the cable in the trench with a 3-inch layer of sand, shovel in 4 inches of dirt and compact the dirt with a 4-by-4. Continue adding 4-inch layers of compacted dirt until the trench is almost full, then replace the sod.

3

Wiring for Communication

A protected path for a stereo signal. A coaxial cable *(bottom)* carries a tiny current from the phonograph cartridge in the background to a nearby amplifier. The braided-wire sheath of the cable shields an inner conductor from electrical interference. To make the connection at the amplifier, the sheath and the conductor are soldered to the shell and to the protruding pin of the plug at center right. For this delicate soldering job, the right tools are a lightweight soldering iron and a roll of rosin-core solder.

A power company provides you with electricity at either 120 or 240 volts—but a modern household needs more choices and a wider range of voltage levels. Routinely, the power company's electricity is changed to higher and lower levels by a device called a transformer. Television sets, fluorescent lamps and many other appliances have built-in transformers that increase, or step up, 120-volt current to levels as high as several thousand volts. In most household wiring, however, transformers reduce voltage. A doorbell or chime, for example, is powered by a transformer that steps down 120-volt power to 20 volts or less. The thermostat that regulates a furnace is supplied by a transformer. And similar step-down transformers are essential components in the control and communication systems described in the following two chapters.

One advantage of low voltage is that it provides enough power for small electrical devices without danger of shock or fire. Because you can use small wires with light insulation and route them practically anywhere inside or outside the house, low-voltage wiring is less expensive and easier to install than stiff, bulky high-voltage cables. Moreover, the big cables are governed by strict provisions of local and national codes that do not apply to low voltage. Many low-voltage connections can be made with a dab of solder *(left)*—a method forbidden in high-voltage wiring.

For the home electrician, low-voltage currents supplied by transformers have a variety of applications. The buzzer circuits of a private telephone system and the voice circuits of an intercom run from transformers. Low voltage activates remote controls for appliances that operate at higher voltages—to switch the 120-volt lights and receptacles in a house, or to signal a 120-volt motor to open and close a garage door. In television and hi-fi systems, specialized transformers control the volume of sound, fit multiple sets of speakers to a single amplifier, power a television antenna rotator and boost the strength of signals from the antenna.

These special-use transformers often require special mounting hardware and connecting devices, but the more common power and signal transformers can be simply plugged into a receptacle or wired to any convenient 120-volt junction box. Since transformers can be damaged by excessive heat, they should not be installed in hot attics or directly above high-wattage lights. Never cover a transformer with insulation or mount it in a confined space: it must be ventilated so that heat will dissipate and it must be accessible so that you can change wiring connections as your system expands. Properly installed, a transformer, which has no moving parts and operates at close to 100 per cent efficiency, should last indefinitely.

Low-voltage Cable: Easy to Connect and Conceal

Wires for circuits of 30 volts or less are generally smaller, lighter and more flexible than regular 120- or 240-volt house wiring, and therefore are very easy to handle. They present little or no hazard, so they are subject to few code regulations. You can install them almost any way you please. Because they may carry not only power but also complex signals for television sets, loudspeakers, intercoms, telephones and remote-control devices, they come in a variety of shapes and colors (below), designed for clear coding to remote connections.

You can hide low-voltage wires in the same places as high-voltage cables so long as the two types are separated by at least 2 inches, but you can also take advantage of routes prohibited for high-voltage wires: along the edge of a carpet, on the surface of a closet wall, down a laundry chute or air duct or along the top of a baseboard. If you cannot conve-

niently route low-voltage wires indoors, you can run them outdoors and then back into the house. And, where their appearance is not obtrusive, you can fasten low-voltage wires directly to walls.

Wherever they are run, low-voltage wires lend themselves to a variety of easy fishing and fastening techniques. For example, a ball of string makes a handy tool for pulling low-voltage wires through joist spaces, and staples or thumbtacks can be used to fasten the wires in place. Low-voltage wires can be connected almost anywhere, without a junction box. They must be connected carefully, of course, to assure trouble-free, safe operation—even at low voltages, sparks from a loose connection could cause a fire.

Smaller sizes of the wire caps and screw terminals used for house wiring will serve for low-voltage connections, although some should be soldered. You can also use a variety of connectors that

can be crimped onto the ends of wires; the insulation on the sleeve of a crimp connector is color-coded for wire sizes—red for Nos. 22 to 18 gauge, blue for Nos. 16 to 14. In addition, several specialized connectors are made for the cables that interconnect hi-fi components and join television sets to antenna systems.

When planning a route for low-voltage wire, keep these common-sense considerations in mind. Do not run wires directly under rugs or carpets or anywhere else they would be damaged by foot traffic or furniture. If you run a route outdoors or in an air duct, use wires rated for resistance to moisture, heat and stress. Be particularly careful to run television and hi-fi cables in gentle curves; sharp bends are temptingly easy to make, but they can alter the electrical properties of the wire and hasten deterioration of the insulation. And never attach high-voltage plugs or receptacles to low-voltage wires.

An Array of Wires and Cables

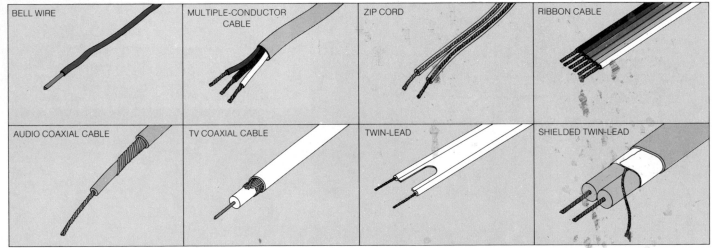

BELL WIRE

MULTIPLE-CONDUCTOR CABLE

ZIP CORD

RIBBON CABLE

AUDIO COAXIAL CABLE

TV COAXIAL CABLE

TWIN-LEAD

SHIELDED TWIN-LEAD

Choosing the right wire. The most commonly used type is called bell wire or hookup wire, either in single strands or in multiple-conductor cables containing from two to several dozen color-coded bell wires that are twisted together and encased in a plastic jacket.

Ordinary lamp cord, known as zip cord since it can be separated—or zipped apart—along a groove in the insulation, is also used on low-voltage circuits. The coding can be by ridges or colored stripes on the insulation or by color or a colored thread in the stranded wire. This example has clear plastic insulation to reveal one

silver-coded wire and one copper-colored. Ribbon or rainbow cable looks like multiple-conductor zip cord and can be zipped open in the same way; its conductors are color-coded strands of bell wire joined with solvent.

For wiring stereo components other than speakers, audio coaxial cable is used. It consists of a series of concentric rings or sheaths: a stranded center conductor, a ring of foam insulation, a second ring of twisted or braided outer conductor and an outer ring of insulation. TV coaxial cable—recommended for connecting TV sets to antennas—is similar, but has a solid center con-

ductor, thicker inner insulation—often plastic foam—and a weatherproof outer insulation. In both types, the braided wire forms a shield to protect the center conductor from interference.

Flat twin-lead was once the universal TV antenna wire, but it is becoming less common because of its susceptibility to interference; it is still commonly used, however, for indoor FM antennas. It is two conductors set about ¼ inch apart in plastic insulation. Shielded twin-lead, an improved version, contains two foam-insulated wires wrapped in foil sheathing and protected by an outer layer of plastic insulation.

Two Techniques for Running Cable

Fastening wires in the open. Use staples, driven by a staple gun with a slip-on wiring attachment *(inset),* to fasten bell wire, jacketed cable or zip cord to joists, baseboards or moldings. Center the staple gun over the wire, straddling it, and hold the gun lightly against the wire so you do not crush insulation. To fasten twin-lead, set the staple gun at right angles to the wires and staple into the insulation between conductors.

Ribbon cable and other flat wires too wide to fit inside staples can be secured with thumbtacks pushed through the insulation between conductors. Use special clips *(page 92)* for coaxial cable. Secure shielded twin-lead and jacketed cables too large for any of the other fasteners with the cable clamps or staples normally used for high-voltage cables.

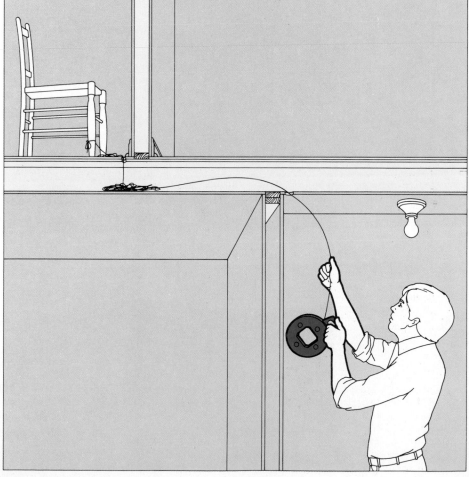

Fishing wires between floors. Next to a baseboard on an upper floor, drill a hole down into the space between the floor joists. Feed a length of heavy string, secured by a nail or furniture leg, into the hole; then, from the lower room or a closet, drill a hole into the same joist space and feed a fish tape through the hole toward the string. Twist the fish tape to snag the string, pull it down through the hole, tie the wire to it and pull the wire to the upper floor.

How to Hide Low-voltage Wires

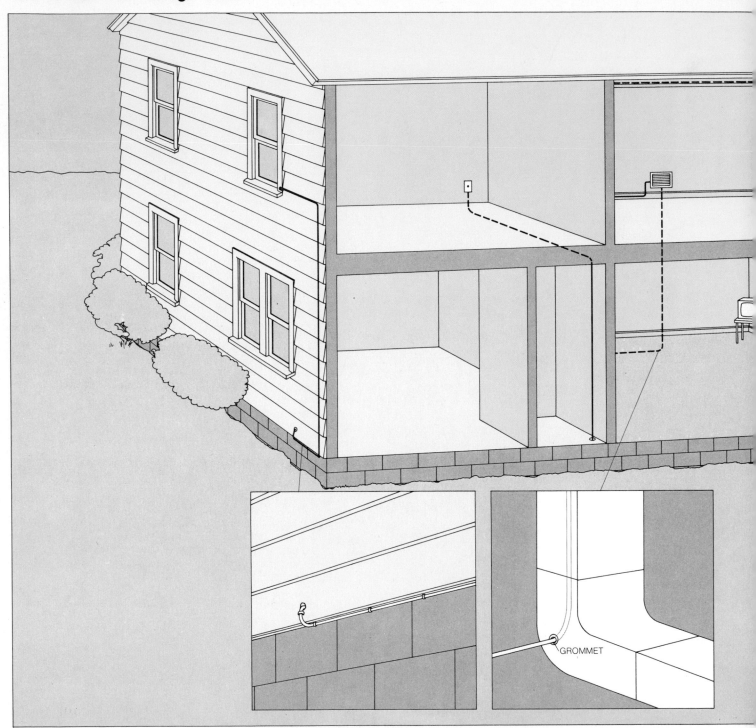

GROMMET

A houseful of wires. This cutaway house contains more low-voltage wires, in more places, than any house is likely to have—but it shows many of the ways such wiring can be completely hidden. Or the way wires can simply be stapled to a wall as inconspicuously as possible.

Outside the house. Drill a small hole through the house wall just above the foundation, run the wire through the hole and staple it along the underside of such sidings as clapboard, plywood sheet or shingles. Drill a second hole in the wall or through a window frame to bring the wire back inside. Caution: use only wire rated for exterior use in an outdoor run and seal each hole with silicone caulking.

Through a duct. Use a fish tape to snake heat-resistant wire with noncombustible insulation through the ducts of a forced-air heating system and run the wire into the duct through a vent or a specially drilled hole. At a drilled hole use a rubber grommet, available at electronics stores, to protect the wire.

At a molding. Tuck the wire into the space between a ceiling and the molding at the top of a wall, securing it at corners with broad-headed thumbtacks. Alternatively, staple wire to the underside of chair-rail molding.

Behind the trim. Use a utility knife to slit the paint along the line where a baseboard, cove molding or doorcasing meets the wall. Pry the trim loose with a chisel and mallet, and run wire in the gap between the wallboard and the floor; in a plaster wall, chisel a groove for the wire. Replace the baseboard carefully, angling any new nails to miss the wire.

At the edge of a carpet. Poke the wire down into the space between the bottom of a wall and the edge of a carpet. If the carpet is too thin to conceal the wire, use a crochet hook to pull the edge of the carpet slightly away from the wall, tuck the wire under the edge and smooth the carpet back into place.

At a doorway. Pry up or unscrew the edging strip that covers the joint between two types of flooring, staple the wire along the edge and replace the strip. Caution: take special care not to drive nails or screws through the wire.

Preparing Cables
for Their Connections

Stripping multiconductor cable. Cut a short lengthwise slit at the end of the sheathing, then hold the slit end of the sheathing—but not the wires—with a pair of pliers, and pull one of the wires out from the slit with a second pair, tearing the sheathing. Snip off the torn sheathing and strip the individual wires.

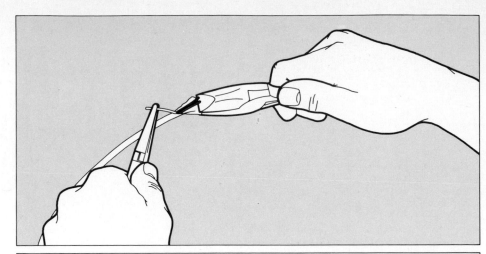

Stripping TV coaxial cable. Using a wire stripper, remove ¾ inch of the outer cable sheathing; then pull the braided shielding wires away from the inner conductor, trim them to ¼ inch and fold the ends back over the edge of the sheathing. Finally, use the 16-gauge setting on the stripper to remove ⅜ inch of the insulation from the inner conductor.

If the cable has a loose foil shielding between the braided conductor and the insulation, peel it off *(inset)*. If foil is bonded to the foam insulation, do not disturb it—use a stripper to remove both foil and foam from the inner conductor.

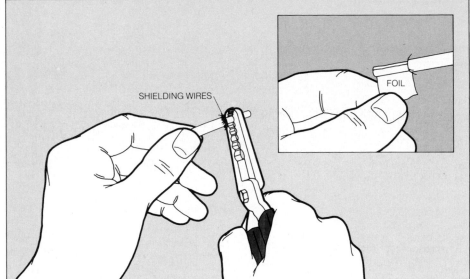

SHIELDING WIRES

FOIL

Crimping Connectors
onto the Cables

Crimp connectors. Strip ¼ inch of insulation from a wire and slip a connector, color-coded for the wire size, over the bared wire end. Use the matching, color-coded crimping die of a multipurpose tool to crush the sleeve of the connector onto the wire.

This method is used to attach lugs for screw terminals, as well as to attach male and female quick connectors *(inset)*.

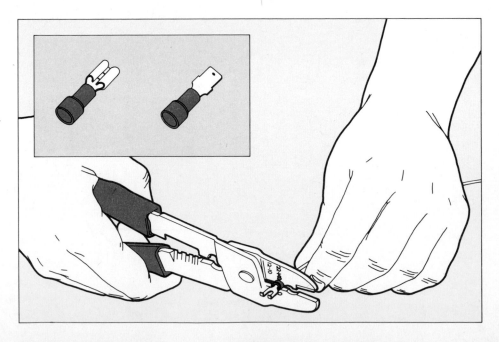

A connector for TV coaxial cable. Slip the metal collar, or ferrule, supplied with an F-59 coaxial connector over the end of RG-59 cable and push the connector over the inner insulation so that the tapered neck of the connector slips under the braided conductor (*below, left*). When the neck is completely covered by insulation and the braiding fits snugly against the back of the connector, slide the ferrule over the neck and crimp the ferrule firmly with long-nose pliers to secure it to the cable.

For audio coaxial cable, solder both inner and outer conductors to a phono pin plug (*page 73*).

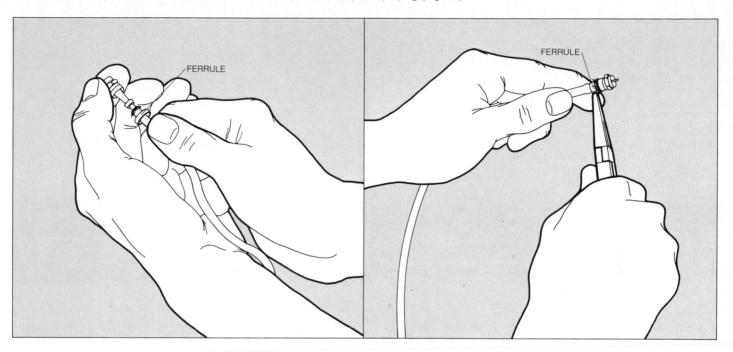

Splicing with a Tight-fitting Tube

Although splices should be avoided—they are insecure unless very carefully made—they can be done fairly readily in low-voltage wiring with a soldered joint protected by heat-shrinkable tubing, available at electronics stores in widths that match most wires.

To prepare two speaker wires—usually zip cord—for splicing, slip a 6-inch length of ¼-inch tubing over one cord and separate about 4 inches of the conductors in each cord. To offset the splices, cut 2 inches from the copper conductor of one cord and from the silver conductor of the other. Strip 1 inch of insulation from all four conductors, then splice the conductors of one cord to the matching conductors of the other by twisting the bared wire ends together and soldering each connection. Slide the tubing over both splices and hold a match just under it, moving the flame rapidly back and forth until all of the tubing has shrunk evenly.

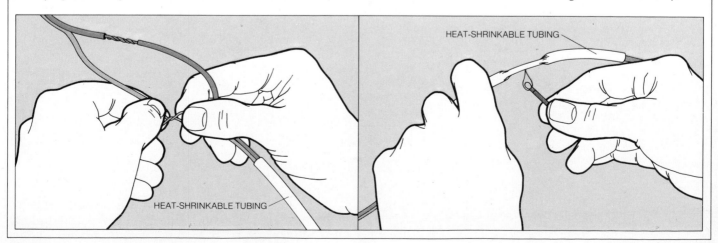

The Simple Art of Soldering

To guarantee mechanically secure, electrically correct connections in many low-voltage circuits, wires must be soldered to flat metal tabs or inside the cylindrical prong of a pin plug. Modern soldering irons and guns help make the job easy. The iron consists of a pencil-thin metal bar containing a concealed heating element and secured in a heatproof handle; a 25- to 50-watt element is adequate for the electrical connections in this book. A soldering gun, generally fitted with an adjustable trigger that controls the temperature of the tip, heats the work more quickly; it is comparatively expensive but worth the extra money if you do a good deal of precision soldering.

Electrical solder, an easy-melting wire of tin and lead with a chemical core to aid bonding, comes in several varieties. Use rosin-core solder for all electrical work—never use acid-core solder, which looks somewhat similar but is designed for plumbing and heavy metal work and can corrode small wires.

The rosin core, called flux, acts as a high-temperature cleaning agent, preventing oxides from forming on metal as it is heated. These oxides, and in fact any dirt, are the cause of most soldering problems. Use a file, a stiff wire brush or a scrap of sandpaper to clean the tip of the iron before you begin to solder. Keep a clean, damp sponge nearby and dab the tip of the iron on it frequently to remove foreign matter. Also thoroughly clean all surfaces that are to be soldered; if possible, bring the surface to a shine.

After dirt, the principal fault in soldering is uneven heating. To avoid this problem, always heat the larger piece in a connection, letting it conduct heat to the smaller. Never heat the solder directly; hot solder dripping onto a cold surface creates a poor bond, or none at all.

Common-sense caution is necessary for safety—the temperature of the tip of a soldering iron is about 850°F. and of molten solder about 700°F. Place a small asbestos pad under the work to prevent burning the work surface. Provide good light and place your work so that you can see the connection clearly. Never leave a plugged-in iron unattended.

Soldering to a Terminal

1 **Tinning the iron.** Heat the soldering iron and dab a small amount of solder directly onto its tip with a brushing motion. Wipe the solder-covered tip clean on a moist cloth or sponge and reapply a small amount of solder.

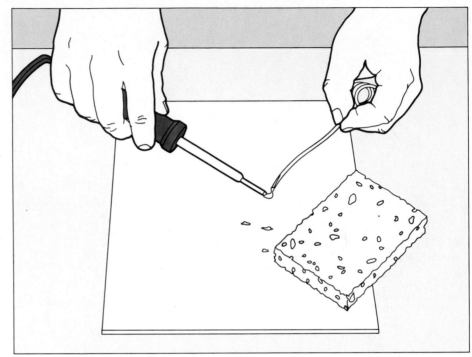

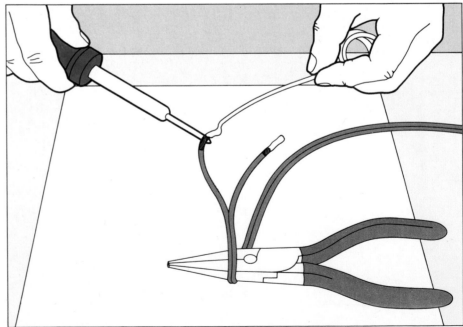

2 **Tinning the leads.** Twist the stripped strands of a wire together and heat the wire with the iron. Apply a small amount of solder to the wire—not the iron. If the wire is hot enough, the solder will melt almost instantly and coat the wire evenly. If the solder does not melt, heat the wire a little longer and try again.

You may find it helpful to hold the wire steady with a weight, such as the pliers shown here.

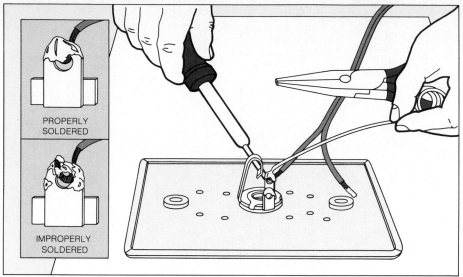

3 Making the connection. Pull the wire through the tab with long-nose pliers, wrap it tightly in a loop around the tab and press the tip of the iron to the underside of the joint. When the joint is heated, touch the solder to the joint and melt just enough solder to cover it. The solder should look rounded and smooth *(top inset)*. If it is uneven and lumpy *(bottom inset)*, reheat the joint; usually reheating will distribute the solder properly and you will not need to add more.

PROPERLY
SOLDERED

IMPROPERLY
SOLDERED

Soldering to a Pin Plug

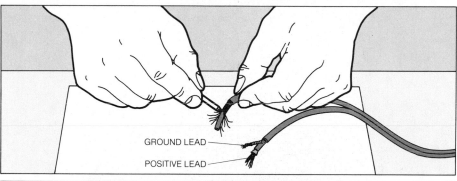

GROUND LEAD

POSITIVE LEAD

1 Preparing the leads. Strip about 1½ inches of the outer insulation and, using the point of a nail, tease apart the braided or spiral ground wire that surrounds an insulated positive lead. Twist the ground wire tightly to form a lead and strip ¾ inch of the positive lead. Tin the soldering iron and both leads *(Steps 1 and 2)* and trim the ground lead to ⅛ inch.

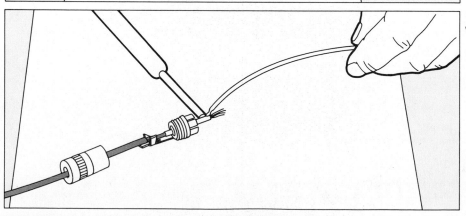

GROUND TERMINAL

2 Preparing the plug. Heat the ground terminal, which also serves as a strain relief, at the back of the plug and melt a drop of solder onto it about ¼ inch from the end. Push the cap of the plug up over the stripped wire, and poke the wire through the plug so that the positive lead protrudes through the end of the prong.

3 Securing the plug. Aligning the plug so that the ground lead lies on the solder dot on the ground terminal, heat the prong of the plug. Apply solder so that it flows down the wire inside and closes the tip opening. Cut off the excess wire and use the hot iron to wipe away any solder that has congealed on the plug. Then place the iron under the ground terminal to melt the dot of solder and bond the ground wire. Pinch the ground terminal around the cable with long-nose pliers and screw on the cap.

Piping High Fidelity All Around the House

Remote-controlled high fidelity systems come in many versions nowadays—and you can pay your money and take your choice. For about the price of a new automobile, you can have a system designed and installed by a professional that permits you to select records, tapes or FM stations from afar and play the same program in every room or different programs in different rooms. Somewhat less versatile but far more affordable is a system that you can install yourself, using the techniques on the following pages. This system begins with the components of any stereo sound system, and incorporates additional components to play a single program simultaneously in as many rooms as you like.

Almost every amplifier has an extra set of terminals for a second pair of speakers at a remote location. Start by adding this extra set of speakers. With bookshelf speakers in a den linked to a living-room stereo system, you can hear the same program in both rooms or listen to the stereo in the den while other members of the family watch television in the living room. The same arrangement would work well with the second set of speakers in a bedroom; in a kitchen, where shelf space is often at a premium, you can get similar results with a pair of speakers recessed in the ceiling.

Adding more than one pair of speakers calls for other components. Some amplifiers have terminals for two sets of remote speakers, but these terminals are deceptive: not even the most expensive and powerful amplifiers can drive more than two sets of speakers simultaneously—normally, one main set and one remote—without a risk of overload and damage. To drive three pairs of speakers at once, you must use a device called a multiple speaker switch box, available at electronics parts stores, to protect the amplifier. And if you want to run four or more pairs, you will need the sound distribution system on pages 79-81.

Separate volume controls for remote speakers (opposite, top) enable you to alter the volume at every location without returning to the amplifier. Ask your dealer for a volume control of the auto-transformer type, install it in a location near the speakers and wire it to the amplifier with a three-conductor cable.

A set of permanently installed speaker outlets will give you an added measure of flexibility—you can plug and unplug speakers to move them from room to room or from indoors to outdoors. The outdoor outlets must be protected by the type of weatherproof enclosure normally used for 120-volt receptacles, but for a permanent outdoor speaker installation, you can wire weatherproof speakers in exactly the same way as indoor speakers, or install recessed speakers (opposite, bottom) in a protected location.

For private indoor listening at a remote location or even in the same room as a television set, use a wall-mounted stereo headphone outlet with its own volume control. Because headphones require less power than speakers, you can connect as many as a dozen headphone outlets to one set of amplifier terminals.

Hookups and Controls for Remote Speakers

Connecting remote speakers. Run coded two-conductor cables—No. 18 for runs to 60 feet, No. 16 for longer runs—between the amplifier and the remote speakers, strip ½ inch of insulation from the end of each conductor and tin the ends (page 72). At the back panel of the amplifier (inset, bottom), connect the conductor from the positive terminal of the left speaker to the amplifier's left positive terminal, identified by a plus sign or red dot. In this example, a cable with one silver wire is used, and the silver wire is restricted to positive connections; however, many coding variations are available. Attach the other conductor to the negative terminal, marked with a minus sign or black dot; repeat the procedure with the right speaker wires. At each speaker, match wires and terminals in the same way (inset, top).

Some speakers and amplifiers have spring-loaded clamps, color-coded red for positive and black for negative, rather than screw terminals; tin wires before inserting them into the clamps. Other models accept spade lugs or RCA phone plugs (page 73).

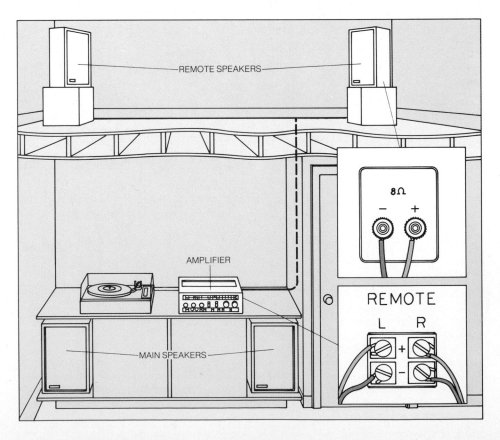

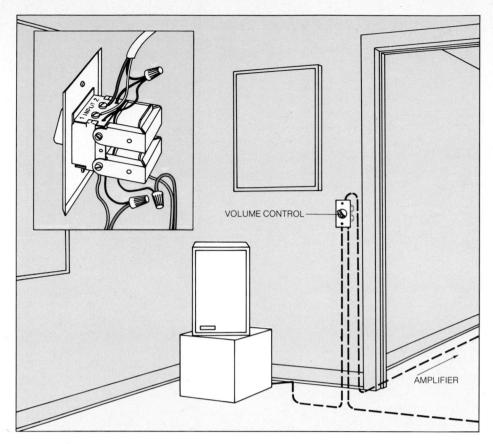

Wiring a remote volume control. Cut a rectangular hole 2 inches by 2½ inches for the volume control; from the hole, run a cable containing red, black and white wires to the amplifier and a coded two-conductor speaker cable to each speaker. At the amplifier, connect the red wire of the three-conductor cable to the right positive terminal, the white wire to the left positive terminal and the black wire to the left negative terminal. Connect the two-conductor cable wires to the speaker terminals as shown opposite.

Working on the control *(inset)* on the side marked INPUT, connect the white wire from the amplifier to input terminal 1 and the red wire to input terminal 2; using a wire cap, connect the black wire to the two black leads attached to the input side of the control. Working on the output side of the control—located at the bottom of the control and not visible in this picture, but identical in numbering and layout to the input side—fasten the positive wire (silver in this example) of the left speaker to terminal 1, the positive of the right speaker to output terminal 2, and the negative wire from each speaker to the corresponding black output lead. Screw the control faceplate to brackets in the wall hole *(page 94, bottom)*.

A Ceiling-mounted Speaker

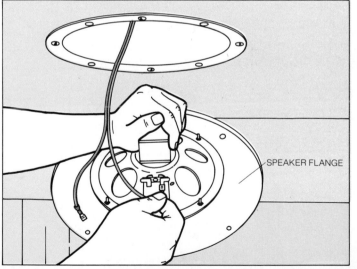

1 **Installing the enclosure.** Between two ceiling joists, cut a hole the width of the speaker enclosure, set the enclosure in the hole and, with a pencil inserted through the flange at its base, mark the positions of the mounting holes on the ceiling. Remove the enclosure, drill pilot holes and install screw anchors in every other pilot hole. Fish two-conductor speaker wire through the enclosure hole, thread it through a knockout and screw the enclosure to the anchors.

2 **Wiring the speaker.** Using a solderless connector *(page 70)*, attach the positive speaker wire (silver in this example) to the positive speaker terminal, indicated by a red dot or a plus sign on the speaker frame; join the other wire to the other terminal. Screw the speaker flange to the open mounting holes in the enclosure flange. Install a second recessed speaker in the same way, and connect both speakers to an amplifier or a remote volume control.

Plugs for Portability

1 **Wiring a speaker plug.** Slip the cap of a ¼-inch phone plug over a No. 18 two-conductor coded cable, cut the copper-colored wire ½ inch shorter than the other, tin the ends of both and attach the copper wire to the long screw terminal of the plug. Fasten the silver-colored wire to the short terminal; then screw the cap onto the plug. Attach the other end of the wire to a speaker (*page 74*).

2 **Wiring a speaker outlet.** At a convenient location for a speaker, drill a 1-inch hole in the wall, fish speaker cable from an amplifier or remote volume control and solder the conductors to the terminals of a speaker jack mounted on a wall plate. Connect the silver-colored wire to the spring contact and the copper-colored wire to the sleeve contact. Insert the jack in the hole and screw the plate to the wall.

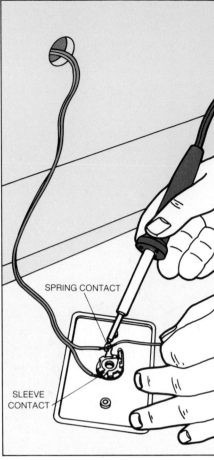

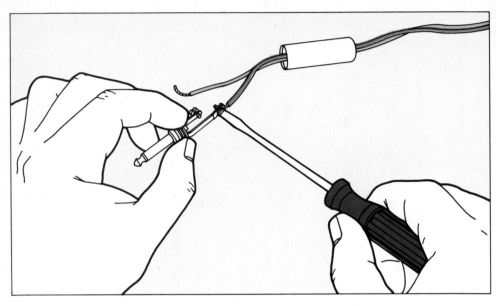

SPRING CONTACT

SLEEVE CONTACT

Earphones for Privacy

AUDIO LEVEL

MIN.

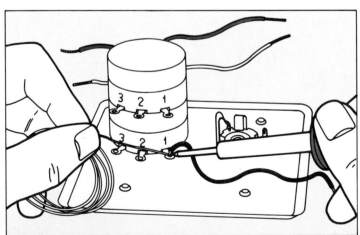

1 **Assembling the wall plate.** A volume control and jack for headphones are mounted by modifying a standard blank double-gang switch plate. Drill ⅜-inch holes midway between the upper and lower screw holes in the plate, and slide the shaft of a stereo L-pad volume control through one hole from the back; slip the marked dial over the shaft, fasten it with a nut and washer and slip on the knob. Fasten a three-conductor stereo headphone jack into the other hole.

2 **Soldering jumpers to the control.** Solder 4-inch jumper wires to three terminals of the volume control: a black wire to the No. 1 terminal next to the mounting plate, a white wire to the No. 2 terminal next to the plate and a red wire to the No. 2 terminal of the volume control.

3 **Linking the control and the jack.** Use a continuity tester to determine the connections between the jack's terminals and its three contacts. With the tester's alligator clip snapped onto a spring contact, touch the probe to the terminals until the bulb lights and label each terminal-and-contact pair. After matching all the contacts and terminals, solder the red jumper from the control to the terminal for the short spring contact, the white jumper to the terminal for the long spring contact and the black jumper to the terminal for the sleeve contact.

4 **Wiring the volume control.** Strip 3 inches of sheathing at one end of a 12-inch length of three-conductor cable, then strip 1½ inches of insulation from the black conductor. Feed the bare wire through the unused No. 1 terminal of the volume control onto the other No. 1 terminal; loop the end of the wire around this terminal and solder the wire at both terminals, taking care not to disconnect the jumper already connected to the No. 1 plate terminal. Solder the white wire to the No. 3 terminal next to the plate and the red wire to the other No. 3 terminal.

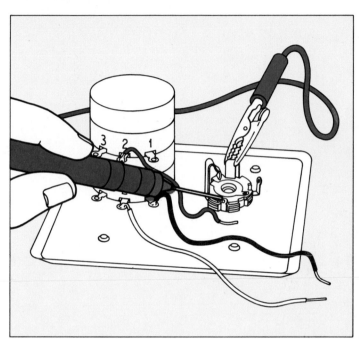

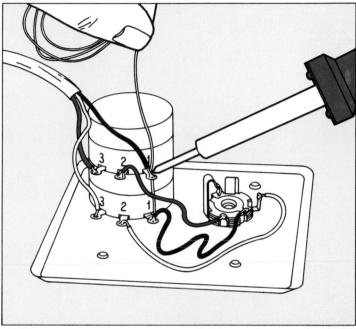

5 **Completing the connections.** At a convenient location near a couch or chair, make a wall opening 2½ inches high and 4 inches wide for the volume control/jack assembly. Fish a three-conductor cable from the amplifier to the hole and use wire caps to connect the wires of this cable to the matching wires of the assembly cable. Screw the assembly into the wall.

At the amplifier (inset), use wire caps to attach 330-ohm, 2-watt resistors to the white and red wires of the three-conductor cable, then fasten the resistors to the left and right positive terminals and the black wire to the left negative terminal. Use separate resistors for each additional headphone outlet.

Taking Hi-Fi Out of Doors

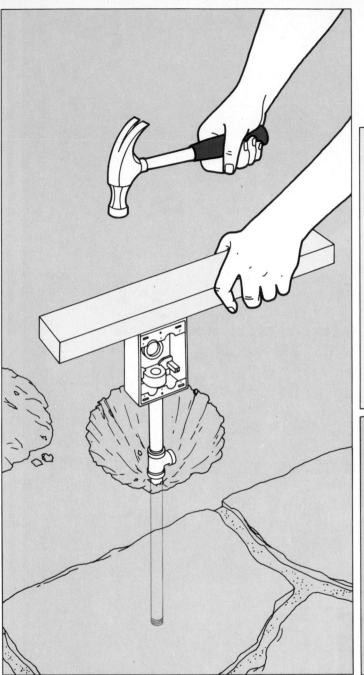

2 Making an adapter plate. To support speaker jacks inside the box, adapt a standard blank switch plate. Trim it to 2 inches by 4 inches with snips and drill ⅜-inch holes, ⅞ inch along the line from each mounting hole. Mount speaker jacks in the holes. Solder 6 inches of white wire to the spring-contact terminal of the upper jack, a similar red wire to the spring-contact terminal of the lower jack and a similar black wire to the remaining terminal on each jack (inset).

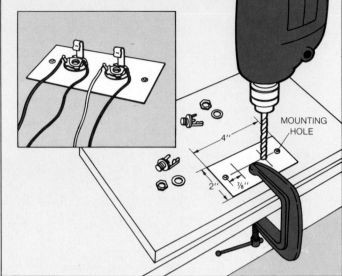

MOUNTING HOLE

4"

2"

⅞"

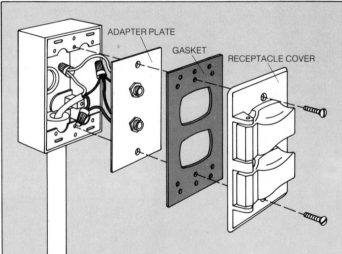

ADAPTER PLATE

GASKET

RECEPTACLE COVER

1 Installing an outlet box. The outlet is assembled from a 6-inch and a 12-inch length of ½-inch steel pipe, a tee fitting and an outdoor electrical box, and simply hammered into a 4-inch hole. Using a scrap of lumber to protect the box, drive it until the tee is at the bottom of the hole.

Dig a trench 4 inches deep between the box and the house and drill a hole through the house wall above the trench. Run No. 16 or 18 three-conductor cable through the hole to the indoor amplifier. Outside, lay the cable in the trench and run it through the tee and up to the box.

3 Mounting the adapter plate. Use wire caps to connect the wires from the speaker jacks to the matching wires of the three-conductor cable installed in Step 1. Make new mounting holes in a standard outdoor receptacle cover and its gasket, using the adapter plate as a template for the drill bit. Screw the three pieces to the box with 1-inch Nos. 6 to 32 screws.

Indoors, connect the three-conductor cable to the amplifier (*page 74*). Outdoors, plug the left speaker into the upper jack and the right speaker into the lower jack.

Wiring for a Multitude of Speakers

With a powerful amplifier, you can operate two main speakers and three or more pairs of remote speakers from it by installing devices called 25-volt line transformers in pairs between the amplifier and each set of remote speakers. Amplifier, speakers and transformers must be matched to one another by the following series of calculations.

To determine the size of the amplifier needed, start with the minimum power requirement, in watts, for each speaker you will use, as given in the manufacturer's specifications. A high-fidelity bookshelf speaker typically requires between 15 and 30 watts; a ceiling speaker may need only five. Add the wattages of all the remote speakers on one channel; the result, which should not exceed 150, is the minimum amplifier power you need.

Line transformers are available with maximum wattage ratings ranging from 5 to 60, increasing in 10- to 30-watt increments; most models also have terminals for intermediate wattages. For example, 30-watt transformers have terminals for 15, 7.5 and 3.75 watts; 20-watt transformers have terminals for 10, 5 and 2.5.

To determine the necessary wattage of a 25-volt transformer, divide the number 75 by the amplifier wattage and multiply the result by the speaker's minimum power requirement. Thus, to match a 25-watt amplifier and a 15-watt speaker, divide 75 by 25 and multiply by 15. The answer is 45, and you must use a transformer with a 45-watt terminal. If the result falls between two terminal ratings, use the terminal with the lower rating.

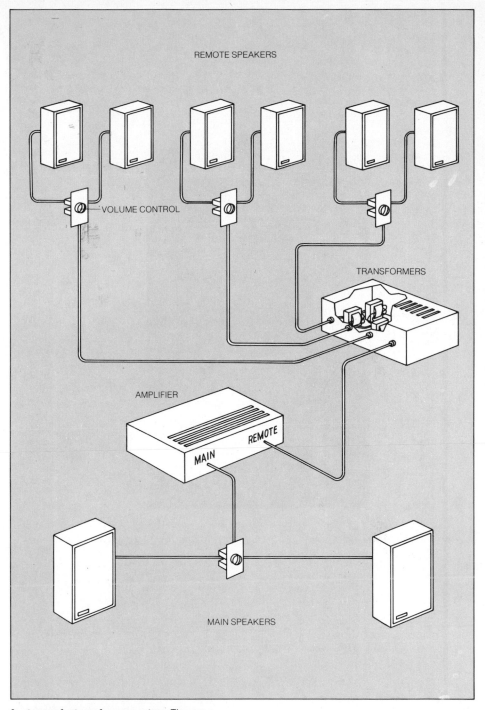

REMOTE SPEAKERS

VOLUME CONTROL

TRANSFORMERS

AMPLIFIER

REMOTE

MAIN

MAIN SPEAKERS

Anatomy of a transformer system. The nerve center of this multiple-speaker system, all connected to one powerful amplifier, is an electronics utility box containing pairs of 25-volt line transformers matched to the speakers and connected to them through individual volume controls by three-conductor cables. In this system the main speakers, too, have an individual volume control, connected to the amplifier by three-conductor cables. The wiring connections inside the box are shown on page 81; the box itself can be mounted on a shelf near the amplifier or in any other convenient location.

Installing the Transformers

1 Wiring the transformers. Solder black wire 12 inches long to the transformer terminal marked PRIMARY COMMON or PRI COM and a similar red lead to the terminal marked for the wattage as determined on page 79. Solder a yellow wire 6 inches long to the terminal marked SPEAKER COMMON or SPKR COM and a similar blue lead to the terminal marked 8 ohms (or 8 Ω). Wire a second transformer in the same way. Then wire the other transformer pairs, matching each terminal to the wattage of the speaker it will feed. Label each transformer with the location and channel of its speaker.

2 Adapting a utility box. Fasten pairs of transformers to the shelf of a utility box with No. 8 sheet-metal screws in ³/₃₂-inch holes.

Drill ⁷/₁₆-inch holes evenly spaced across the middle of the faceplate and fit them with ⁷/₁₆-inch grommets; you will need grommeted holes for an amplifier cable and for one speaker cable from each pair of transformers.

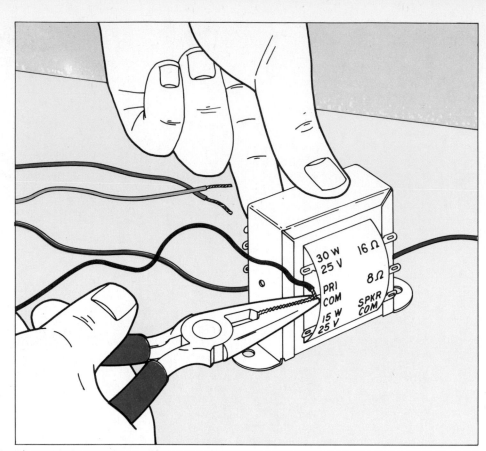

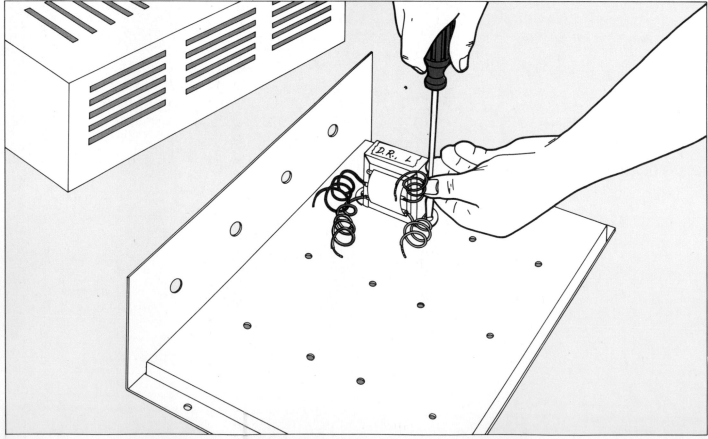

3 **Connecting the amplifier leads.** Run a No. 18 or 16 three-conductor cable coded red, white and black from the amplifier's remote speaker terminals into a grommeted hole at one end of the utility box. With wire caps, join the red wires of all the left-channel transformers to the white wire from the amplifier, the red wires of all the right-channel transformers to the red wire, and the black transformer leads to the black wire.

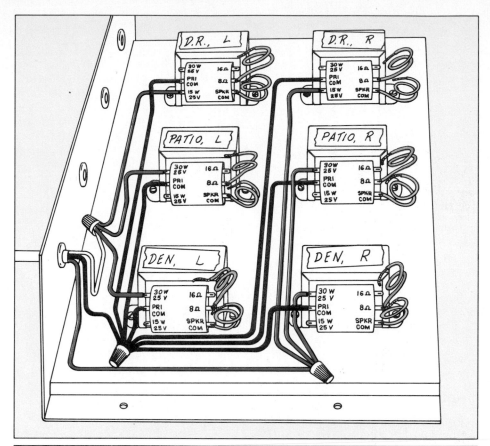

4 **Connecting the volume controls.** Run a No. 18 or 16 three-conductor cable coded red, white and black from each remote volume control to the transformer box and feed the cable through the grommet closest to the appropriate transformer. At each pair of transformers, attach the white cable lead to the blue lead of the left channel unit, the red cable lead to the blue lead of the right channel unit, and the black cable lead to the yellow leads from both transformers.

Install the cover on the transformer box and complete the wiring connections at the amplifier, volume controls and speakers (*pages 74-75*).

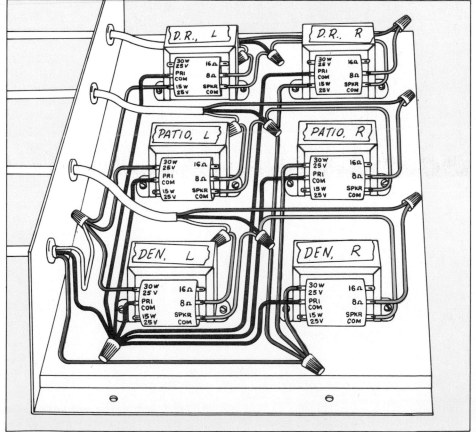

Room-to-room Messages by Intercom or Telephone

Modern intercommunication systems—intercoms—were originally developed for military use in airplanes and tanks, but today they enable anyone in the family to question a visitor outside the entrance door; talk with others in a workshop, a basement or a bedroom; monitor the sounds in a nursery; or even pipe music to every room in the house.

A number of different types of home intercoms are available. You can, for example, simply order the telephone company to install extensions with switch and buzzer buttons so that you can communicate between rooms. But a wide variety of equipment is sold by electronics-parts stores to enable you to provide much the same service yourself. The most popular system, described on the following pages, is powered by a low-voltage transformer connected to the house electricity. Simpler to install—it just plugs into wall outlets—is the so-called wireless intercom, which is actually a short-range FM radio transmitter sending signals over the house wiring to receivers plugged into outlets in other rooms; however, because it transmits on a maximum of three different frequencies, it cannot carry messages to more than three separate receivers. At the other end of the scale is a private telephone exchange for your home that you can assemble from readily obtainable telephone equipment *(pages 86-87)*.

The low-voltage setup is the most widely used because it combines versatility with simplicity. It generally has one master station linked by switches to a number of remote stations, each consisting of a speaker-microphone fitted to a mounting, or rough-in, frame. The system generally includes an AM-FM radio at the master station, fed either by an indoor FM antenna provided with the set or by an outdoor antenna you can install yourself *(pages 88-93)*.

The master station switches can be set to send messages or music to the remote stations, to listen in on them or to hold them on stand-by for calls. At the remote stations, microphones can be turned off for privacy while the speaker is left on to receive calls from the master station. And along with words and music, some systems can carry the sounds of door chimes, alarm clocks and burglar alarms.

Locate the master and remote stations for your convenience, but keep a few basic rules in mind. Choose a central location, such as a kitchen or family room, for the master station. Avoid placing any room station on an exterior wall because in mounting it there you may damage the vapor barrier on the inner surface of insulation. A door speaker, however, must generally go in an insulated wall; use the trick of wiring through studs shown on page 14 to avoid piercing the vapor barrier there. And to prevent sound interference or feedback within the system, do not mount stations back to back on opposite sides of the same wall or facing each other through open doors or archways. Locate the transformer in an inconspicuous place—a closet or basement, perhaps—where the cable runs so that connections to the master station and a 120-volt outlet box will be simple.

You can run the low-voltage wires connecting the units on the surface of interior walls *(pages 68-69)* or concealed inside walls as shown here. Before cutting the wall holes for station frames, check as best you can for obstructions such as heating and cooling ducts, major plumbing lines and fire stops that would interfere with cable runs. Take special care to identify electrical lines: intercom wires must be at least 12 inches away from 120-volt wires or telephone cables.

Tag each end of the cables with numbered pieces of adhesive paper, available at electrical-parts suppliers, and fish each remote station wire to a central location in a closet, attic or basement. This wiring method is likely to create longer routes for some cables, but it saves time—you can fish all the cables to the master station in a single operation and will have less trouble in locating the cables for repairs or alterations.

To drain small electrical charges that might interfere with intercom or radio reception, ground the master station rough-in frame to the nearest cold-water pipe with a No. 14 copper grounding wire. An 18-gauge, two-conductor cable carries power to the master station from the low-voltage transformer terminals.

In the typical system shown on these pages, two-conductor cables connect the master to the door stations and three-conductor cables join the master and remote stations; more sophisticated systems have cables containing as many as six color-coded conductors. Most intercom manufacturers will honor their warranties only if you use their cables to make the connections between stations.

Installing an Intercom System

1 **Cutting station holes.** At a convenient location for the master station, use the back of the rough-in frame as a template to mark the outline of the frame between two studs. Cut away the wall surface within the outline. Make similar rough-in holes for the indoor remote stations alongside studs, aligning each frame so that the mounting holes on one side fall directly over a stud *(inset)*. Do not install any of the frames at this point. Stations for use from outside the house, as at a door, are mounted as explained in Step 2.

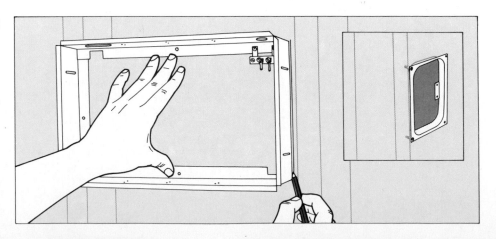

2 Preparing a front-door station. Along the outer edge of the doubled studs at one side of the doorway—generally an inch from the edge of the trim—make a vertical mark on the siding at the station height you have chosen. Set the rough-in frame with its side edge on this mark and its top edge against the bottom of a clapboard, shingle or brick. Mark the outline of the frame on the siding, and cut through the siding and sheathing along the outline.

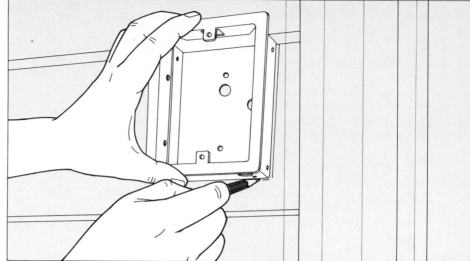

3 Installing the transformer. At a convenient location accessible to a 120-volt circuit in a basement or closet, mount an octagonal 4-inch box. Thread the high-voltage leads of the transformer through a knockout, insert the transformer's clamp assembly into the knockout and tighten the setscrew to fasten the transformer in place. After cutting off power to the 120-volt circuit to be tapped, run No. 14 plastic-sheathed cable from its most accessible box to the transformer box. Connect the hot and neutral cable wires to the black and white transformer leads and screw the cable ground wire to the box. Cover the box with a metal plate.

Attach a grounding clamp (*page 45*) to a cold-water pipe convenient to the transformer.

Using techniques described on pages 12-14 and 67, fish the following wires to the master station: a No. 14 copper wire from the grounding clamp, a No. 18 two-conductor cable from the transformer, and the cable specified by the manufacturer from each remote speaker. If you have an outdoor antenna, connect it to the intercom radio with a signal splitter (*page 94*); otherwise install an indoor FM antenna (*below*).

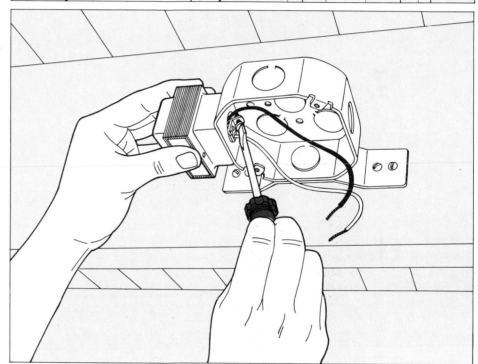

An Antenna Mount for an Intercom Radio

Attic rafters provide the best mounting surface for an indoor FM antenna, generally supplied with an intercom kit, but if you have no attic you can mount it in a closet. In either location, the insulated wires that make up the receiving element should be mounted to form a horizontal U shape. The elements are connected to the antenna terminals at the intercom master station by the flat twin-lead wire sometimes used for a TV antenna lead-in or with coaxial cable.

In the attic, nail a board at least 6 feet long between two rafters and staple the center section of the antenna, about half as long as the entire receiving element, to the underside of the board. Run the antenna ends at right angles to the board, stapling them to the undersides of adjoining rafters. If you are mounting the antenna in a closet, tape the wires in a U shape on the walls.

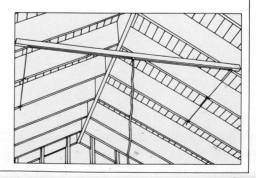

4 **Connecting the power and ground wires.** At the transformer, strip the insulation from the leads of the low-voltage power cable and connect them to the transformer's low-voltage terminals; either wire can go to either terminal. Fasten the ground wire to the grounding clamp.

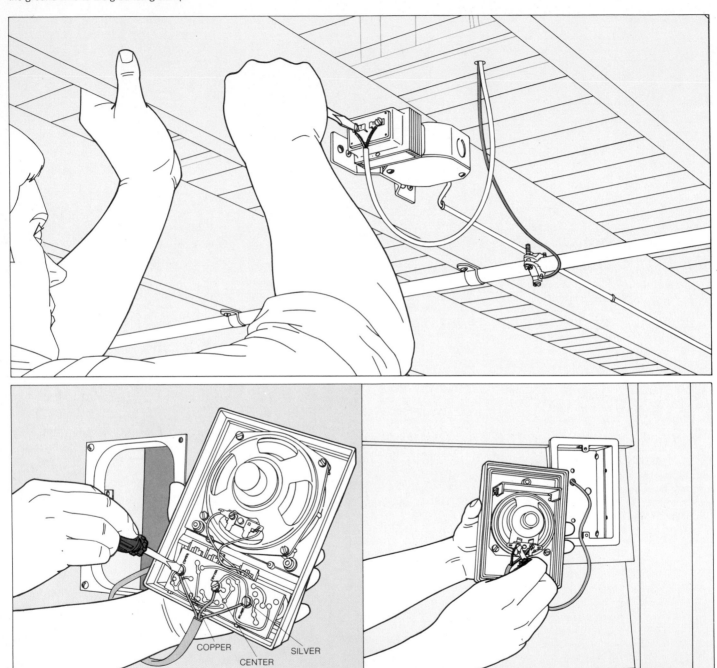

COPPER
CENTER
SILVER

5 **Wiring a remote station.** At an indoor station *(left)*, fasten the rough-in frame to the adjoining stud and attach the cable wires to the back of the station, following the manufacturer's color code. To read the code, strip off a bit of insulation to see the color of the bared wire. In this typical installation, copper- and silver-colored wires are fastened to the matching screw terminals; the remaining wire goes to the center ter-

minal. Push extra cable out through the hole in the frame and fasten the station to the frame.

At an outdoor station *(right)*, thread the cable through the hole in the back or bottom of the rough-in frame, and screw the frame to the doorframe studs. Attach the wires following the color-coding, mount the station in the frame and caulk the outer edges of the frame.

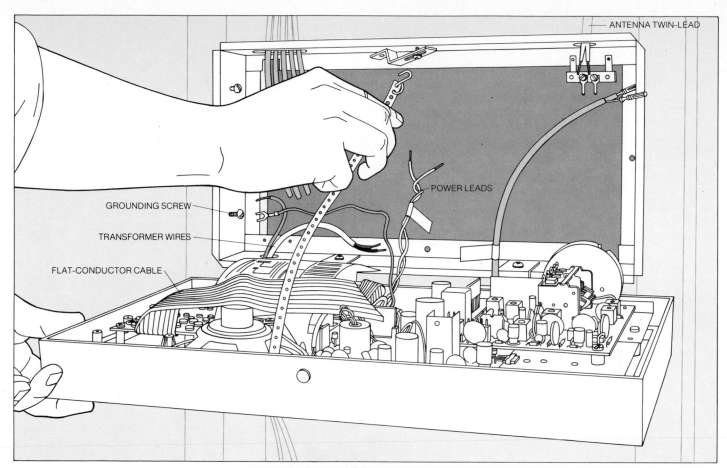

ANTENNA TWIN-LEAD

POWER LEADS

GROUNDING SCREW

TRANSFORMER WIRES

FLAT-CONDUCTOR CABLE

6 **Mounting the master station.** Feed all wires into the frame, screw the frame to the studs, hang the master station to the hinges and hook on the support strap. Join the transformer wires and power leads, and secure the ground wires from the grounding clamp and master station to a grounding screw on the frame. Connect the antenna twin-lead to the screw terminals just inside the frame. The prominent flat-conductor cable is prewired by the manufacturer.

7 **Linking master and remote stations.** Following the manufacturer's instructions, connect the cable from each remote station to screw terminals in the master station panel. To be sure of making the right connections, assign a number to each station, tag the station cable with a numbered piece of adhesive paper and match the cables to the numbered terminals of the panel. Most models, including the one shown here, have specially marked terminals for door stations and accessories such as chimes.

Raise the master station into place, fasten it to the rough-in frame with the screws or locking pin provided by the manufacturer and restore power to the circuit powering the transformer.

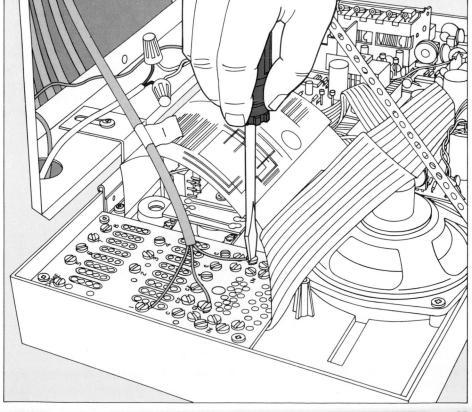

Assembling a Private Phone Network

The most flexible communication is provided by the telephone. Unlike intercoms, it is not under the control of a master station, from which or to which every call must be made. With parts that are available at electronics stores, a simple party-line phone system can be built into a home. The miniature system has its limitations. It uses individual buzzers, rather than dials, and the party-line conversations can be heard by anyone who picks up a telephone. But otherwise it is as flexible as its larger cousins.

The heart of the home system is a central exchange, a set of interconnected strips linking all buzzers and handsets. The setup is powered by a 6-volt battery for voices and a 10-volt transformer for the buzzers. (You need the buzzers because the bell inside a telephone cannot be used without a type of current that is impractical to generate at home.)

The first phone serves as a sound generator for the entire system, and it must be off the hook at all times; in practice, the switch buttons are removed from the cradle so that the system will not go dead when this telephone is hung up. Install this phone in a room where it will not pick up distracting noises.

For clarity, telephones in the accompanying pictures and captions are identified by letters, terminals by numbers.

2 **Connecting the button pad and buzzer.** Connect terminals 3 and 4 of the block to the two buzzer terminals with jumper wires. Open the button pad, which comes with a cord containing one wire for each button and one common to all. Connect the common wire, yellow here, to terminal 5 of the block. Connect the wires from the buttons in the alphabetical order of the other telephones to the terminals above 5 in numerical order. In the example shown at right, there is of course no button for telephone A; the button for telephone B is connected to terminal 6, and the button for C to terminal 7. At telephone B, the wiring would be A button to terminal 6, C to 7; at C, A to 6, B to 7.

In larger systems the connections follow the same logical pattern.

1 **Connecting the telephone.** Loosen the screws in the base of the telephone, remove the cover and examine the attached wall cord, which will contain three or four wires, generally including a green one connected to the terminal marked L1. Connect the L1 wire to terminal 1 of a terminal block mounted on a nearby baseboard. Connect the wire from L2, red in this example, to terminal 2. Snip off the other wires at the cable sheathing.

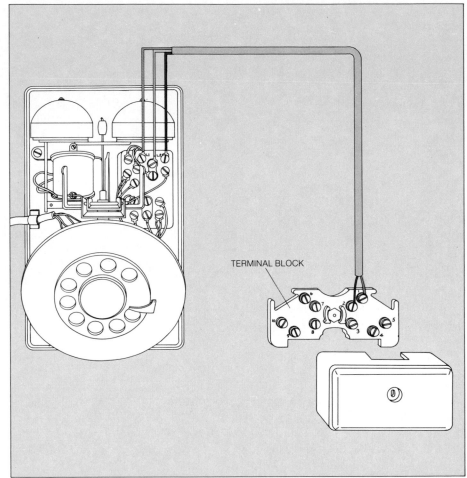

TERMINAL BLOCK

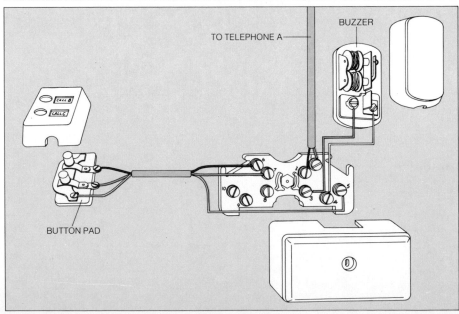

TO TELEPHONE A

BUZZER

BUTTON PAD

3 **Connecting the cables.** Run a No. 22 cable containing at least four more color-coded wires than the number of telephones in your system from each terminal block to a terminal strip at the central exchange, normally in an out-of-the-way spot. Use the color codes to connect each numbered terminal on a block to the matching numbered terminal on the left side of a terminal strip. This scheme requires two wires to run from each terminal of the block, a practice permissible in low-voltage circuits; simply wrap the second wire around the terminal with the first and take care to tighten the screw so that it holds both wires securely.

4 **Wiring the exchange.** At the central exchange, follow the wiring pattern shown below, making all terminal-strip connections on the right-hand sides of the strips. Connect A1 to the positive terminal of a 6-volt battery with a short jumper wire. Link C2 to B2, and B2 to the negative terminal of the battery. Link all the No. 3 terminals together and run a wire from them to one output terminal of the transformer; link all No. 5 terminals to the other output terminal. Working from strip to strip, make the following three-terminal linkages with jumper wires: A2, B1 and C1; A4, B6 and C6; A6, B4 and C7; and A7, B7 and C4.

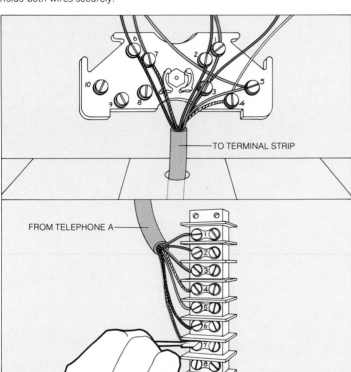

TO TERMINAL STRIP

FROM TELEPHONE A

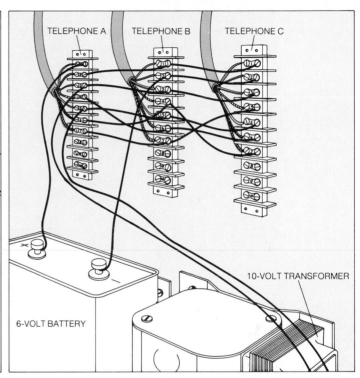

TELEPHONE A   TELEPHONE B   TELEPHONE C

10-VOLT TRANSFORMER

6-VOLT BATTERY

Wiring for Additional Phones

D	E	F
C1 - D1	D1 - E1	E1 - F1
C2 - D2	D2 - E2	E2 - F2
C3 - D3	D3 - E3	E3 - F3
C4 - D8	D4 - E9	E4 - F10
C5 - D5	D5 - E5	E5 - F5
C6 - D6	D6 - E6	E6 - F6
C7 - D7	D7 - E7	E7 - F7
A8 - B8 - C8 - D4	D8 - E8	E8 - F8
	A9 - B9 - C9 - D9 - E4	E9 - F9
		A10 - B10 - C10 - D10 - E10 - F4

Expanding the system. To create a system with more than three telephones, wire additional telephone stations as in Steps 1 and 2, and run cables from these to new terminal strips, one for each station. Wire the terminal strip for the fourth telephone, telephone D, to the terminal strip for telephone C. Then connect E to D, and so on, as shown in the table at left.

For a system of more than six telephones, you will need terminal strips with more than 10 pairs of terminals. Continue to expand the central exchange, one strip at a time, making the connections indicated in the tables.

Mounting a Rooftop TV Antenna

An indoor "rabbit ear" antenna often provides adequate TV reception for a black-and-white set. But a color set, which uses more of the TV signal and is more sensitive to uneven antenna reception, will almost always work better with an outdoor antenna. You can install the antenna yourself, following the instructions on the following pages. If you already have an outdoor installation the instructions can help you to improve it by replacing flat lead-in wires with shielded coaxial cable or by protecting the antenna against lightning damage as required by the National Electrical Code.

Antennas come in a variety of shapes and sizes *(below)*. Some shapes pick up a single channel, others pick up VHF (channels 2 to 13), UHF (channels 14 to 83) or FM radio channels, and still others receive VHF, UHF and FM. Whatever its shape, an antenna is sized according to its distance from the transmitting stations in its area. It is sold with either mileage-range figures or such ratings as local, suburban, fringe and far fringe, and its cost increases with its distance rating. Before buying an antenna, consult your dealer on any unusual features of your situation: a location in a valley or behind a large building may call for a larger antenna or a higher mast. The size of the antenna you buy may also depend on whether you plan to install a distribution system *(pages 94-95)* to feed several receivers.

Because almost all antennas are directional—they work best when the front of the antenna points toward a transmitting station—determine the location of the television transmitting towers in your vicinity. You can get this information from your dealer, from the stations themselves or from a local office of the Federal Communications Commission. If you find that transmitters are scattered in your area, you can improve reception with an electrically powered rotator that turns the antenna by remote control *(pages 96-99)*.

In general, the higher the antenna the stronger the signal. But there are practical limitations on antenna height: masts for chimney or wall mounting are generally available only in 5- and 10-foot lengths and taller telescoping masts, up to 50 feet tall, require special hardware and supporting guy wires. If your roof is partially

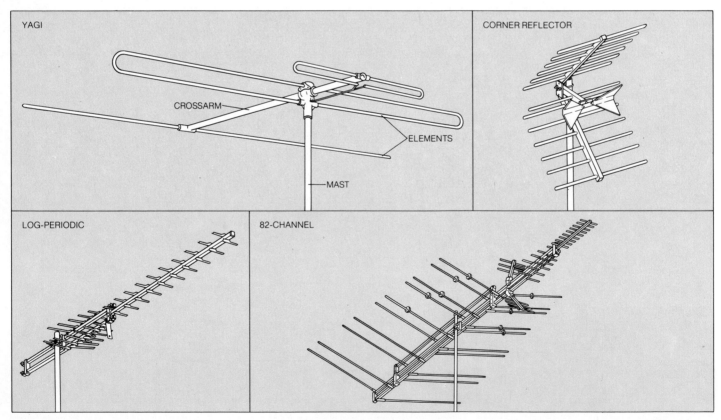

The anatomy of antennas. All four of the antennas above are made of weather-resistant aluminum and have receiving elements arranged on a central crossarm, or boom. It is the shape and arrangement of the elements that indicate the antenna's function. The Yagi antenna, named for its Japanese inventor, uses two looped elements and a straight reflector element to pick up VHF (channels 2 to 13) and the FM radio band. In the corner reflector, a wedge-shaped array of reflectors replaces the straight Yagi reflector and an element in the shape of a bow tie replaces the Yagi loops to form a compact UHF-only antenna. A more sophisticated UHF design called a log-periodic antenna uses a series of successively larger elements that act as reflectors for each other to strengthen the signal. The large VHF-FM-UHF model called an 82-channel antenna combines a log-periodic UHF section, a corner reflector and a log-periodic VHF and FM radio section. The plastic disks on the VHF elements are electronic dividers that enable each element to receive the wavelengths of several different channels.

blocked from the transmitter by a building or hillside, place the antenna on the end of the roof farthest from the obstructions. Never mount an antenna directly above or below electric lines and, whenever possible, separate it from the wires by at least the length of the mast.

Fastening an antenna to a chimney is generally the easiest approach and, when the chimney is the highest point of the roof, the best—but make sure that the chimney is structurally sound and repoint deteriorated mortar joints before attaching the mast. Alternately, you can mount an antenna on the edge of the roof or the side of the house. If you do not have a metal roof or foil-backed insulation between the rafters, you may be able to conceal an ungainly VHF antenna—often some 7 feet wide and 8 feet long—by mounting it in the attic *(page 93)*. UHF

reception is impaired by any roof, especially on rainy days, and attic mounting of a UHF antenna is unadvisable.

Use 75-ohm, shielded coaxial cable for the down-lead from the antenna to the receiver—the flat unshielded twin-lead commonly found in older installations does not stand up well to weather and is prone to electrical losses that can cancel out the antenna signal altogether. Because most television antennas and receivers are still designed with twin-lead terminals, you will probably have to connect a small transformer at each end of the coaxial cable to match the cable to the antenna and receiver.

Keep cable runs as short as possible: signals lose strength as they travel through a wire. To avoid damaging the cable's signal-carrying ability, avoid sharp bends and do not kink, dent or crush the

cable. Outdoors, route the cable at least 2 feet away from electrical power lines; indoors, maintain at least a 2-inch separation between the cable and house wiring. If you have a lightning-rod system, keep the down-lead at least 6 feet from the system conductors and have a professional connect the mast to the system; otherwise, ground the mast as shown on page 91, Step 2. Protect all down-leads and rotator wires with grounding devices or lightning arresters *(page 92, Step 4)*.

Before mounting the antenna, do as much assembly work as you can on the ground. Position ladders as far as possible from electrical service wires and be sure that neither the antenna elements nor the mast will touch overhead wires as you carry the assembly up to the roof. On the roof, use a ladder supported by roof hooks over the peak.

Braces for the Mast

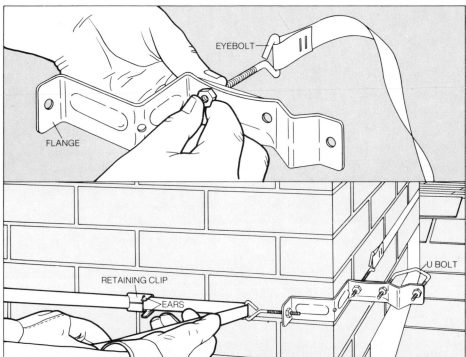

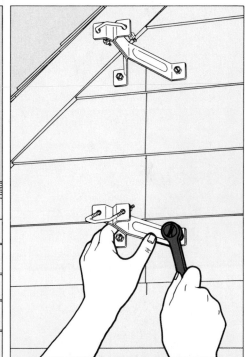

A chimney mount. Working on the ground, assemble a pair of mounting straps and brackets. Push the eyebolt at the end of each strap through the hole at the middle of a bracket and secure the bolt with a few turns of a nut *(top)*. Secure a loose eyebolt to the flange at one end of each bracket and a U bolt to the other end; be sure that the nuts in the bolts are on the same side of the bracket.

On the roof, loop one of the straps around the chimney a foot or so above the roof. Slip a

retaining clip over the end of the strap, thread the strap through the second eyebolt and pull the strap tight to cinch the bracket to a chimney corner *(bottom)*. Feed the end of the strap through the retaining clip, then fold the strap back over the clip toward the bracket and hammer the ends of the clip down over the strap to hold it securely. Cut off the excess strap, then tighten the nuts on the eyebolts. Install the second strap and its bracket farther up the chimney, at least one third of the mast length above the first.

A wall mount. Working on the ground, install U bolts loosely in a pair of wall brackets, then draw a plumb line on the wall at the point you have chosen for the antenna; center the brackets over the plumb line, separated by at least one third of the mast length, and fasten the brackets to the siding and sheathing with lag bolts. For a wall under a wide roof projection, you may need adjustable brackets, which can secure a mast up to 18 inches out from the wall.

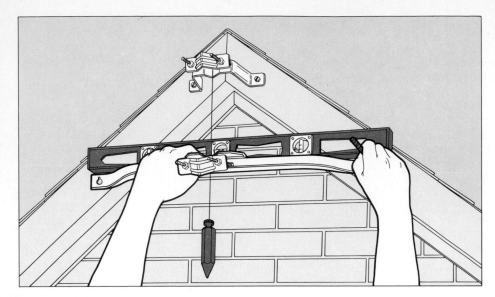

A gable mount. Center and level the upper bracket on the rake boards immediately below the ridge and fasten the bracket with lag screws; then hang a plumb line from the center of this bracket and align the lower bracket for installation using the plumb line and a level. The antenna mast should be no longer than three times the distance between the brackets.

Assembling the Antenna

1 Installing a transformer. Working on the ground, loosen the wing nuts of the two connection terminals on the cross member of the antenna, and slip the spade connectors of a weatherproof matching transformer between the large terminal washers (inset). Tighten the wing nuts onto the small lock washers and coat each terminal with a weatherproofing compound such as silicone grease or petroleum jelly.

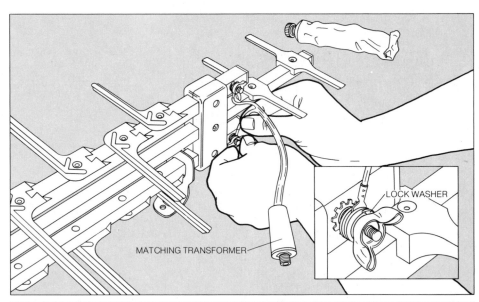

MATCHING TRANSFORMER

LOCK WASHER

2 Connecting the down-lead. Thread the end of the down-lead through any guides on the cross member and slide the rubber boot supplied with the matching transformer over the end of the coaxial cable. Attach an F-type coaxial connector (page 71) to the cable end. Force weatherproofing compound into the weatherproofing boot (right), slide the boot over the connector and screw the connector tightly to the threaded fitting on the transformer.

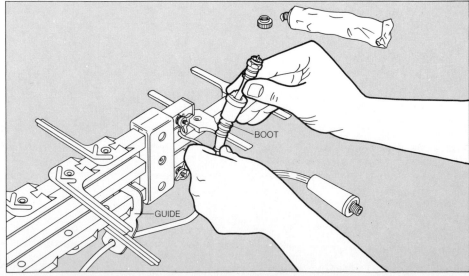

BOOT

GUIDE

3 **Attaching the antenna to the mast.** Wearing gloves, unfold the antenna elements until they lock into their fully open position, then loosen the antenna clamp and slide the mast between the jaws of the clamp until the top of the mast protrudes a few inches above the antenna. If you are also installing a rotator (*pages 96-99*), fasten it to the mast before mounting the antenna. Tape the down-lead and the rotator wires to the mast temporarily.

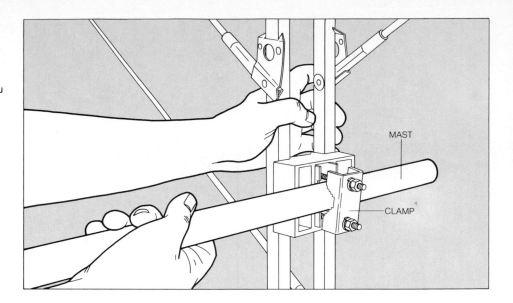

MAST

CLAMP

Completing the Job

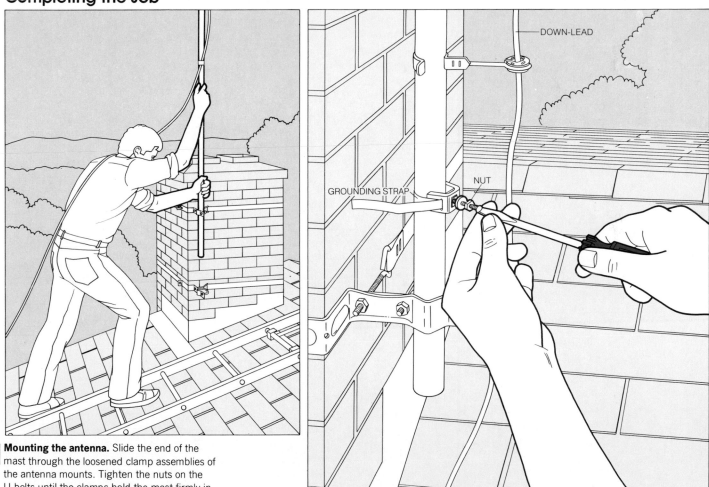

DOWN-LEAD

GROUNDING STRAP

NUT

1 **Mounting the antenna.** Slide the end of the mast through the loosened clamp assemblies of the antenna mounts. Tighten the nuts on the U bolts until the clamps hold the mast firmly in place without crushing it.

Remove the tape from the down-lead and attach standoff insulators to the mast. Thread the cable through the insulators, leaving a loop of wire about 1 foot wide beneath the antenna.

2 **Adding a ground wire.** Screw a grounding strap to the bottom section of the mast and attach an antenna grounding wire, no smaller than No. 8 aluminum or No. 10 copper, under the nut and washer on the grounding-strap screw.

3 Running the down-lead. Secure the down-lead to the side of the building at 4-foot intervals with special coaxial cable clips (*right*), good for masonry as well as wood, or with nail or screw-type standoff insulators. To protect a down-lead that runs across a roof from moisture and abrasion that could affect the signal quality, suspend it by twisting it around a wire stretched from the antenna mast to a standoff at the edge of the roof.

Fasten the ground.wire and the antenna rotator wire (if any) to the side of the house with cable clips or standoff insulators. You can run coaxial cable and rotator wire together in a special insulator (*page 98*), but ground wires should always run separately from the down-lead.

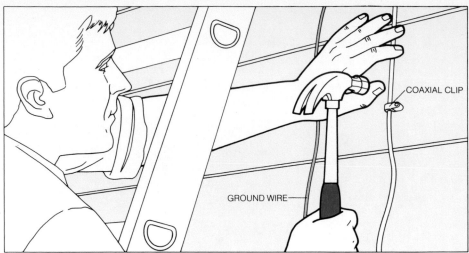

COAXIAL CLIP

GROUND WIRE

4 Grounding the down-lead. Near the point where the cable will enter the house, screw a coaxial grounding block to the wall. Cut the cable at this point, fasten coaxial connectors (*page 71*) to each of the cut ends and screw the connectors to the grounding-block terminals. Insert the end of a grounding wire in the ground-wire hole and tighten the setscrew. Attach the loose ends of the grounding wires from the mast and the grounding block to an outdoor ground rod (*page 46*) or a grounding clamp on a cold-water pipe (*page 45*).

If you use twin-lead rather than coaxial cable or install a rotator, protect the leads with a grounding device called an antenna discharger (*page 98*), available at electrical-supply stores.

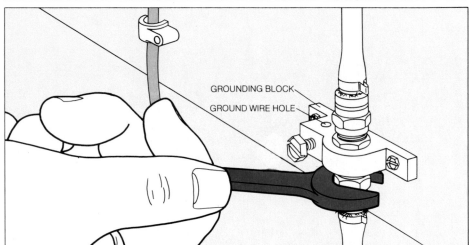

GROUNDING BLOCK

GROUND WIRE HOLE

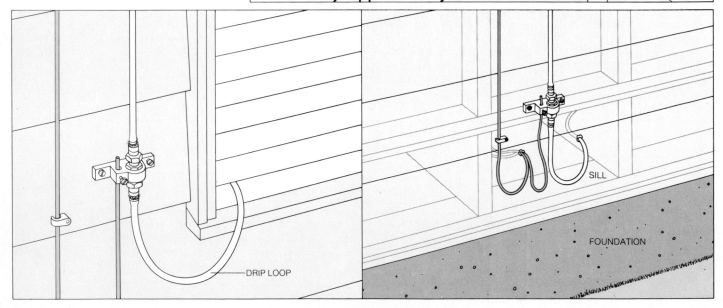

DRIP LOOP

SILL

FOUNDATION

5 Bringing the down-lead indoors. Route the coaxial cable through the bottom louver of a gable vent (*left*), through the siding 4 or 5 inches above the sill on the foundation wall (*right*), or through a channel or hole in a window frame.

Outdoors, form a drip loop and caulk any holes you make. Indoors, run the wire to the television or splitter location, using the techniques on pages 66-69 and keeping the cable at least 2 inches from current-carrying lines.

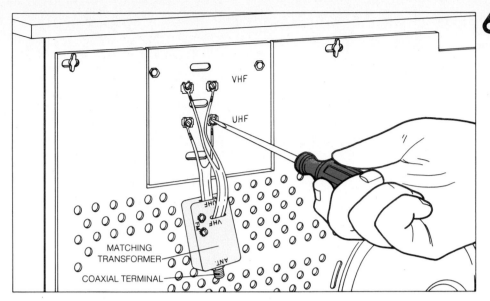

6 **Hooking up the receiver.** Screw the UHF and VHF leads of a matching transformer with separate UHF and VHF outputs to the corresponding terminals on the back of the television. Attach a coaxial connector to the end of the down-lead and screw the down-lead cable to the transformer terminal. Mount the transformer on the back panel of the receiver, using the adhesive supplied by the manufacturer. Transformer terminals are often provided for a twin-lead connection to an FM radio.

VHF

UHF

MATCHING TRANSFORMER

COAXIAL TERMINAL

Concealing an Antenna in an Attic

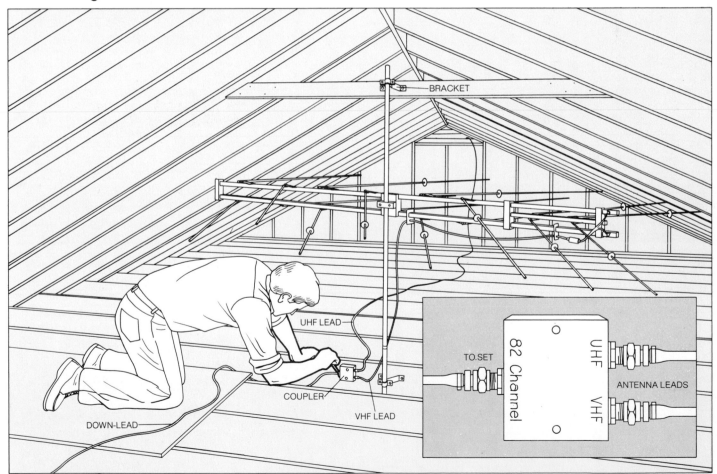

BRACKET

UHF LEAD

COUPLER

VHF LEAD

DOWN-LEAD

TO SET

82 Channel

UHF

VHF

ANTENNA LEADS

Combining the signals. Use a UHF-VHF coupler to join a VHF antenna mounted in an attic with a UHF antenna mounted outdoors. For the attic mounting, attach brackets (*page 89*) to a joist and to a collar beam or a horizontal 2-by-4 nailed between rafters directly above the joist. Clamp the antenna to the mast, hold the mast vertical and aim the antenna front toward nearby TV transmitters; then, without changing the direction of the antenna, fasten the mast to the brackets. Attach the coupler (*inset*) to a joist and thread the coaxial leads from each antenna to the corresponding coupler terminals. Run a down-lead inside the house from the output terminal of the coupler to the receiver or a splitter.

Routing the Signal to Several Sets

For a large family or a large house, an antenna distribution system multiplies the value of an outside antenna. From a single antenna and down-lead, the distribution system can send signals to several TV and FM sets or to several outlets for a portable set. And it is far easier to install than the antenna that feeds it.

If your house is served by a community antenna and a cable, check with the cable company before you put in a system; for a nominal charge, some companies will install a system for you—even to the point of snaking wires inside your walls. Otherwise, install your own system, fitted to your special needs. The simplest system consists of a two-way signal splitter, usually located in a basement, feeding the signal from a large antenna to a pair of sets upstairs. The leads that carry the signal can be brought up through the floor; or, for a neater job, cables can be fished through the walls to permanent wall outlets. At each of the outlets, a distribution lead from the signal splitter goes to a coaxial connector on the back of a wall plate; a short coaxial lead carries the signal to the set from a connector on the front of the plate.

A four-way splitter feeds four sets, but in a location more than 15 miles from a transmitter, a single antenna may not be able to supply four TV receivers. (Even two sets may be too many in fringe areas.) The solution for the problem is a TV signal amplifier that boosts the strength of the signal. Amplifiers are available for VHF, UHF or 82-channel amplification and can create a signal strong enough to feed as many as 16 outlets. The model you choose may have a built-in signal splitter or it may come with a single output terminal that must be hooked to a separate splitter.

Unfortunately an amplifier can create new problems: with its signal amplified, a strong local TV or FM station may overpower weaker stations. Here, the solution consists of traps—devices that reduce the strength of troublesome signals and, in some models, can also eliminate CB-radio interference.

Installing a signal splitter. In an inconspicuous location such as a basement, attic or closet, screw a two- or four-way splitter to a joist or other structural member and attach the coaxial down-lead to the input terminal. Run coaxial leads from the output terminals to a matching transformer at each receiver (*page 93, Step 6*) or to a wall outlet (*bottom*). Cap each unused output terminal with a DC blocking terminator (*inset*), available in antenna or electronics stores.

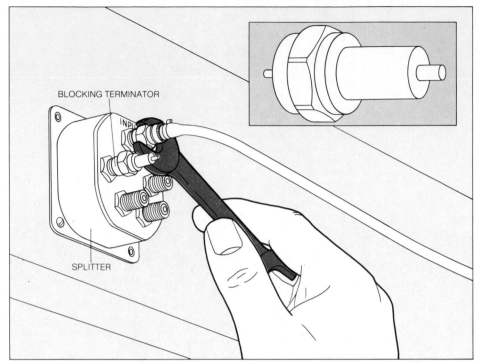

BLOCKING TERMINATOR

INPUT

SPLITTER

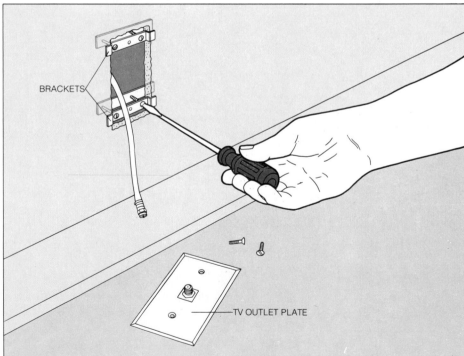

BRACKETS

TV OUTLET PLATE

Mounting a coaxial wall outlet. Fish coaxial cable from the splitter to a hole 1¾ inches wide and 4 inches high cut through the wall at the outlet location. Screw the front and back straps of the outlet brackets loosely together, slip the back straps behind the wall with their central mounting holes 3¾ inches apart, and tighten the screws to clamp the brackets in place. Attach the cable to the connector on the back of the wall plate and screw the plate to the brackets.

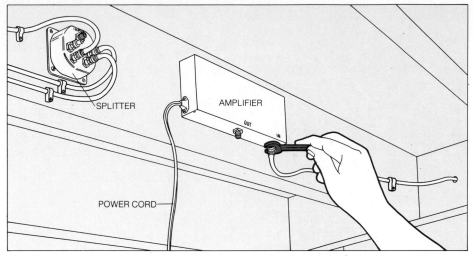

Adding an amplifier. Mount a signal amplifier near the splitter and at a point where its power cord can be plugged or wired into 120-volt house current. Attach the antenna down-lead to the input terminal of the amplifier with a cable connector, and run coaxial cable from the amplifier output to the input terminal of a signal splitter or the antenna terminal of a receiver.

SPLITTER

AMPLIFIER

OUT

IN

POWER CORD

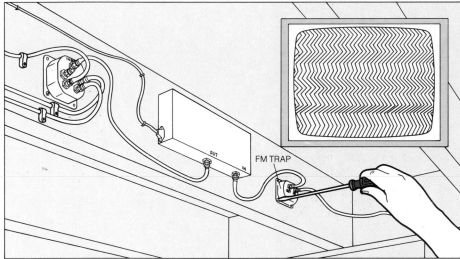

FM TRAP

OUT

IN

Correcting overloads. To eliminate interference from a strong FM signal, often indicated by a herringbone pattern on the TV screen *(inset)* or a completely scrambled picture, wire an FM trap between the antenna down-lead and the amplifier input. You can still run the signal antenna to an FM radio by installing a TV/FM band separator between the antenna and the trap, and by running a cable from the separator to the FM set.

To eliminate an overload from a local TV station, indicated by a negative TV image or by a series of vertical black bars that drift across the screen, wire a single-channel VHF trap for the offending channel into the down-lead exactly as you would an FM trap.

A Testing Tool for a Pro

The field-strength meter at right, which costs as much as a medium-sized color TV set and is not normally available for rent, is no tool for amateurs. But a professional troubleshooter of antenna distribution systems is likely to call upon it as often as an electrician calls upon his voltage tester. Rather than relying upon his eyes and ears to gauge the quality of a TV signal, the pro uses the meter to separate the signal of each TV channel into picture and sound components and to measure the strengths of the separated signals in units called decibels and millivolts.

From the meter readings, he can pinpoint defective antennas, leads or connectors, identify and measure overloads

caused by interference, and determine when a system needs an amplifier. With an antenna down-lead connected to its input terminal, the meter becomes, in effect, a battery-powered TV set with a meter dial in place of the picture tube. The operator sets the selector switch to VHF or UHF, adjusts the meter with the zero-set dial, and tunes to a channel with the numbered dials at the left of the panel for a reading at the dial at right. A headphone jack permits him to listen to the sound signal. When an especially strong signal—a sign of overload—sends the meter dial off the scale, he switches in a series of attenuators *(lower right)* to weaken the signal by a predetermined amount.

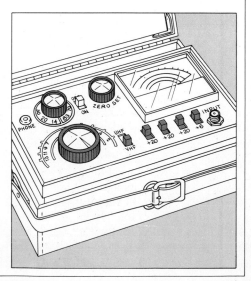

A Rotator to Aim the Antenna

Most television antennas work best when they face a transmitter squarely—that is, when the front of the boom (normally, the end that supports the short elements) points toward the TV station. Receptivity falls off and colors wash out if the antenna points away from the signal source—a swing of 30° will spoil any signal; a smaller swing will spoil reception if you use a complex antenna or tune your TV to a high-number channel.

With an antenna rotator, you can always turn the boom of the antenna in the right direction, sweeping the horizon for incoming signals just as the beam of a turning searchlight picks out distant objects. In locations between cities, the antenna can swing to capture signals from transmitters that lie in opposite directions. In fringe areas it can sharpshoot stations up to 100 miles away with an accuracy of 5° or less—a degree of precision that can make the difference between a clear picture and a snowy one. When a TV picture is spoiled by ghosting (a double image created by a combination of direct signals and signals that are reflected off nearby hillsides or large buildings), the rotator can turn the antenna away from the ghost or, sometimes, deliberately choose it. And as UHF channels between 14 and 68 increasingly go on the air, rotators will become even more valuable, for UHF transmitters are likely to be widely scattered.

A typical rotator consists of two parts: a control box, placed near the set; and a rotor, a motor mounted on the antenna mast in a weatherproof housing. The control activates a transformer that turns the 120-volt current from a wall plug into 24-volt current; from the box a lightweight cable carries the low-voltage current to the motor through circuits that turn the transformer off after the antenna has turned to a new position. To protect the entire assembly, a grounded lightning arrester, called a discharger, is installed on the rotator cable (Step 6).

If your antenna has a flat, twin-lead connection, switch to a coaxial lead-in cable (page 90, Step 2) when you install a rotator: the cable will complement the rotator in improving reception.

Operating an antenna rotator. With the rotator at rest, the control-knob arrow and the direction indicator of this typical control box line up with the point on the compass rim that matches the direction of the antenna. When the viewer turns the knob to another compass point, a lamp will light the dial and the rotor will turn the antenna. The direction indicator will move as well; when it lines up with the new position of the arrow, the rotor and the lamp will switch off. In the rotator shown here the rotor transmits the orientation to the indicator by means of a clicking relay; in another common type, a control-box motor synchronized to the rotor turns the indicator. Because continuous rotation would twist the antenna lead-in until it snapped, the rotor and control knob reach a stop point after a maximum of one full revolution in either direction. In the relay system shown here the stop point can be moved (Step 8) to face away from the sector of the compass that includes frequently viewed channels (inset). In most synchronous-motor systems, the antenna is installed facing north at its stop point; if this direction lies between much-used channels, reinstall the antenna so that it faces south at the stop point and relabel the compass points on the control panel to match the directions of the antenna.

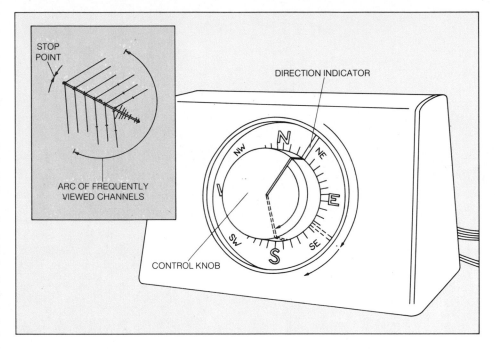

STOP POINT

ARC OF FREQUENTLY VIEWED CHANNELS

DIRECTION INDICATOR

CONTROL KNOB

Installing the Rotator

1 Connecting the rotor. Thread the rotator cable through both the slots and the gasket of the connection-panel cover, strip and tin the ends of the cable wires (page 72) and connect the wires to the terminal screws. This cable contains four wires; other rotator models have three- or five-wire cables, but the installation procedure is always the same. Connect the wires to the most convenient terminals. Note the terminal number of the wire—always at one edge of the cable—that has silver strands and the terminal numbers of the other wires in sequence to the other edge of the cable. The cable wires and their terminal numbers must match on the control box and wall plates (Step 7). Install the gasket and cover plate.

Disconnect the antenna and mast and bring them down from the roof. Hacksaw through the mast about 18 inches below the antenna.

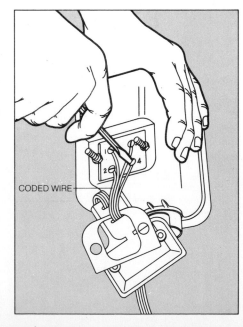

CODED WIRE

2 **Mounting the rotor.** Insert the setscrews provided with the rotator through the clamp plate and screw them loosely to their nuts; then bolt the lower section of the mast between the clamp plate and the rotor, with the cut end against the rotor flange. Tighten the setscrews.

3 **Fastening the antenna to the rotor.** Slide the upper section of the mast down through the rotor shaft and, just above and below the rotor, tighten the U bolts enough to keep the mast from slipping in the shaft. Remount the mast and antenna assembly on the roof.

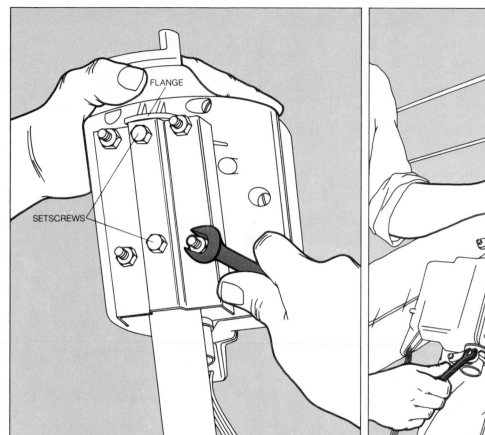

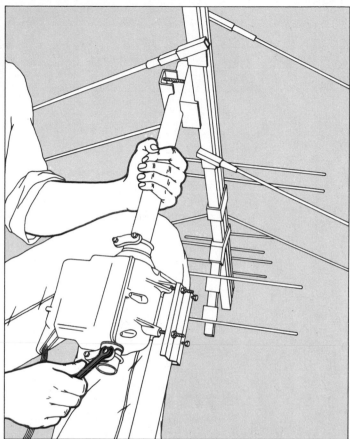

4 **Orienting the antenna.** Loosen the U bolts and turn the antenna until the end of the boom with the longest elements points toward the most frequently viewed TV stations, then tighten the U bolts until the pin in the yoke bites into the mast. Reconnect the antenna lead-in.

If your rotator has a motorized direction indicator rather than the relay type shown in these pages, point the front of the antenna north or south, as directed in the instruction sheet supplied by the manufacturer.

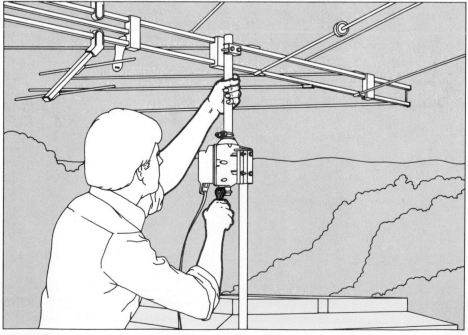

5 **Leading the wires down.** Mount a double standoff insulator on the mast a foot below the rotor. Pop the plastic disk from the outer ring of the standoff; thread the antenna lead-in through the disk and press the disk back into the ring; then form a slack lead-in loop between the standoff and the lead-in guide on the boom long enough for the antenna to turn 360° without binding. With pliers, tighten the ring around the disk to secure the lead-in. Turn the other inner standoff disk to a point where its slot lines up with the gap in the ring, slide the rotator cable into the slot and turn the disk to close the slot.

6 **Installing a discharger.** Remove the cover of a rotator static-and-lightning discharger, fasten the base of the discharger to the roof or wall of the house near the entry point of the rotator cable and run the cable through the discharger's central channel. Replace the cover. Ground the discharger (*page 92, Step 4*) from its ground screw. Finally, run the cable of the rotator into the house through the antenna lead-in opening.

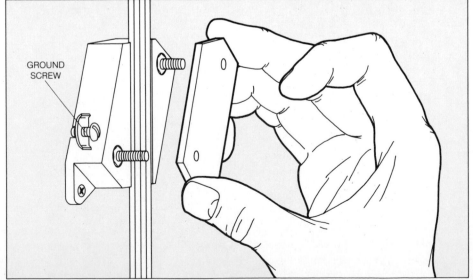

GROUND
SCREW

7 **Connections to wall plates.** For a control box that will not be moved, wire the rotator cable directly to the terminals at the bottom of the box, matching the tinned wires to the same terminal numbers as the rotor *(Step 1)*. If you have several outlets for antenna connections *(pages 94-95)*, use wall plates fitted with receptacles for the cable. Run the cable from plate to plate, making connections at screw terminals *(below, left)* or soldering lugs. Trace the terminal-number sequence of the cable through to the holes in the face of the receptacle and, in a matching pattern, connect a cable to the prongs of the plug *(right)*. Tin the wires at the other end of the plug cable and screw them to the terminals of the control box.

You can now plug the control box into any of the outlets. Do not plug in additional control boxes unless the rotator you are installing will accept them and do not turn the control knob while moving the box between outlets.

(pages 94-95)

(page 97, Step 2)

When a Mast Needs Extra Support

A standard antenna rotator works well for most antennas—but not for all. If you need an especially high antenna or live in an area of frequent high winds, an optional bracket-and-brace assembly called a thrust-bearing can support a 5-foot rotating mast and help the rotor to withstand strong sideways forces. The bearing, fastened to the mast by a clamp plate identical to that of the rotor *(page 97, Step 2)*, may be mounted beneath the rotor, as in the model below, or above it. In either case, set the rotor and bearing from 8 to 18 inches apart and align their clamps in exactly the same direction. Tape the antenna lead-in to the top of the rotating mast and run it down to the standoff insulator in several loose turns around the mast, so that it can wind or unwind freely as the antenna rotates.

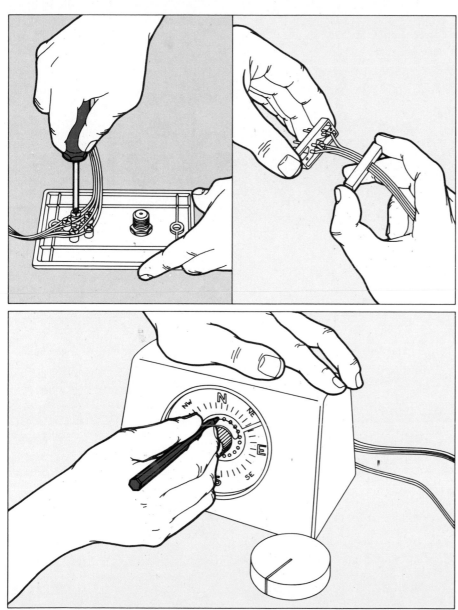

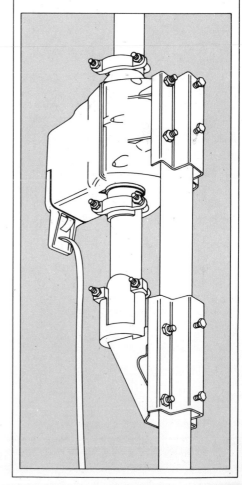

8 **Aligning the direction indicator.** Pull off the control knob and, using a pencil, turn the disk behind it until the arrow points in the same direction as the antenna. Replace the knob, with its arrow lined up to the disk arrow.

High winds and long use occasionally force the antenna and control out of alignment. Test the alignment of a relay-operated control occasionally by turning it to the stop point in one direction, then in the other. If the dial lamp remains lighted at the stop point, press the button in the bottom of the box repeatedly until the lamp turns off. Controls with motorized direction indicators can be realigned by turning the knob to the limit in both directions.

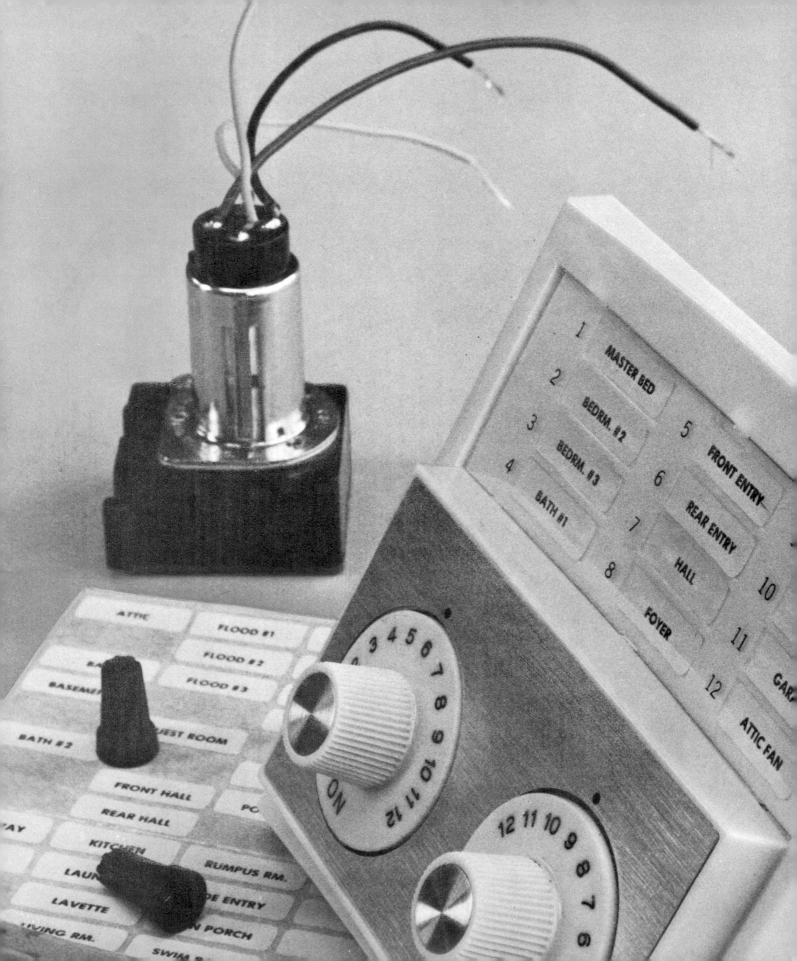

4 The Wizardry of Remote Control

Master and slave. To turn a fixture on from the remote-control master switch at lower right, you consult the directory panel for the fixture's number, turn the ON dial to that number and press the knob inward for a moment. A pulse of low-voltage current goes to a cylindrical relay, an electrically actuated switch installed in the fixture's outlet box, and the relay supplies 120-volt current to a light or receptacle. The miniature wire caps at left make fast and easy connections in the low-voltage part of the circuit.

In 1879, when Thomas Edison lit a bank of electric lights at Menlo Park in the first public display of his new invention, he demonstrated more than the marvel of the lights themselves. At that moment he also introduced to the world the concept of a switch that turns lights on and off from a distance—a convenience unknown to users of candles and kerosene lamps. The demonstration of the light was also a demonstration of remote control.

Today it is simple to go a big step beyond Edison and, using low-voltage wiring and devices called relays, put the electrical equipment in your home under the kind of elaborately convenient control exercised by an astronaut in a space capsule. Any combination of lights and appliances can be made to go on or off by buttons placed in as many different locations as you like. A single switch can control many devices, while each device is still controllable separately. You can make some things go on when others go off. And you can automate the system with motor-operated master switches, timers and light-sensitive cells, so you never have to touch a button at all.

This electrical wizardry is made possible by the relay, which is simply a switch operated by another switch. The relay—one for each device to be controlled—switches on or off the 120-volt current to the device. The relay itself is actuated by 24-volt current from a manually or automatically operated push-button switch—and one relay can be controlled from any number of buttons. In a garage door opener, for example, a touch to a wall button sends low-voltage current to a relay in the mechanism that turns on the motor to move the door. When the movement is nearly completed, another low-voltage switch triggered by the door itself sends a signal to turn the motor off and resets the circuitry to move the door in the opposite direction on the next command. That command may come from the same wall button, from a key button outside the garage or from a radio receiver mounted near the motor. The receiver in turn gets its command from still another switch, in a hand-held transmitter that sends a signal from a car.

Because relays operate on momentary pulses of 24-volt current, they have all the advantages of low-voltage wiring. The components of a circuit are inexpensive small-gauge wires, tiny touch switches and compact master switches. Since the wires present little danger of shock or fire even if short-circuited, they can be run without junction boxes along any convenient and inconspicuous route. These same characteristics make low voltage safe and effective for another application, the lighting of outdoor spaces. Such circuits bring Edison's invention to the garden in wires that can be strung anywhere and connected without the complexities of high-voltage wiring.

Versatile Switching with Low-voltage Relays

When it comes to switching, conventional 120-volt wiring has limitations. Installing a single switch is easy enough, but adding three-way and four-way switches grows cumbersome—and because the full current flows through the switches, they must be fairly close to the fixture to minimize voltage drop.

Low-voltage switching overcomes all of these limitations. With it you can add any number of switches to any number of fixtures, and place the switches wherever you please. You can, for example, install light switches beside every entrance to a room, or turn on a path of light through stairs and halls or from a house to a garage, with the touch of a single switch. You can turn off all the house lights from your bedside—and instantly turn them all back on if you hear a prowler. From an upstairs switch you can start the breakfast coffee perking downstairs in the kitchen. With selective switching from a remote-control panel you can change the mood and decor of any room in the house (pages 64F-64G).

The key element in this kind of wiring is the relay, a magnetically operated switch that obeys commands from other, smaller switches. The relay itself is installed in an outlet box—for example, in the junction box of a ceiling light fixture, where it connects and disconnects 120-volt current to the light. It gets its orders from the other switches in the form of brief "on" and "off" pulses—and because the commands consist of small surges of current, the rest of the system runs on 24 volts, supplied by a doorbell-type transformer. The low voltage eliminates any chance of shock from the command switches, and the small command currents need only small-gauge wires that, under most codes, can be run without junction boxes and even outside of walls by the techniques shown on pages 66-69.

For the command switches, sensitive seesaw-style devices operated by the tap of a finger replace the toggle switches of conventional wiring. The more sophisticated and imaginative switching systems, using rotating master switches, hand or motor operated, send patterns of "on" and "off" commands to a number of relays at once. You can also switch with automatic timers and photoelectric cells.

Because low-voltage switching is flexible—you can put in an additional switch to an existing relay whenever you wish—you need not plan the circuitry all at once and can work on the system a little at a time. You might start, for example, by installing a relay in a hall light and connecting several switches to it; later, if you put in a master switch, you can simply run a wire from it to the relay.

Installing the parts of a system—command switches, relays, a transformer and wiring—follows a simple, standardized procedure. In remodeling work you will generally make wall holes big enough for one or more command switches. After wiring a switch, you simply snap it into a strap and screw the strap to the face of the wall. In new construction the switch straps screw to punched metal plates called plaster rings, which are nailed to studs and then concealed by wallboard. The same straps also fit into conventional junction boxes. And for masonry or for wiring that need not be concealed, there are compact strapless switches that can simply be nailed to the outside of walls.

Relays slip into outlet boxes even more easily than switches fit onto walls. A barrel containing the low-voltage magnets slips through a knockout hole in a box and snaps into place with the 120-volt terminals inside the box and the 24-volt connections outside. If an existing junction box is too shallow for the relay, you must replace it. To control a receptacle with a relay, you must gang the receptacle box to an identical box (page 105, Step 3). If the momentary low hum of an operating relay is audible and objectionable, install the relay in a junction box that is out of hearing range and run a standard high-voltage cable between it and the receptacle or fixture.

Install the transformer at a location that is out of sight, accessible to a 120-volt circuit and convenient to the bulk of the relays and switches. Wiring the circuit consists mostly of running two- or three-conductor cables of No. 18 stranded copper wires, connected with wire caps and coded by the colors of their insulation; the color code of the wires on the following pages is that of the most commonly used system. Always leave at least 6 inches of extra cable at each switch and relay so that you can, if you wish, pull these circuit elements out for inspection or replacement.

Complete an installation by removing the old high-voltage switches and joining the black wires that lead to them (page 105, Step 4). If you want to use their boxes for low-voltage switches, use No. 18 wires with an insulation value of 600 volts and special low-voltage switches equipped with screw terminals.

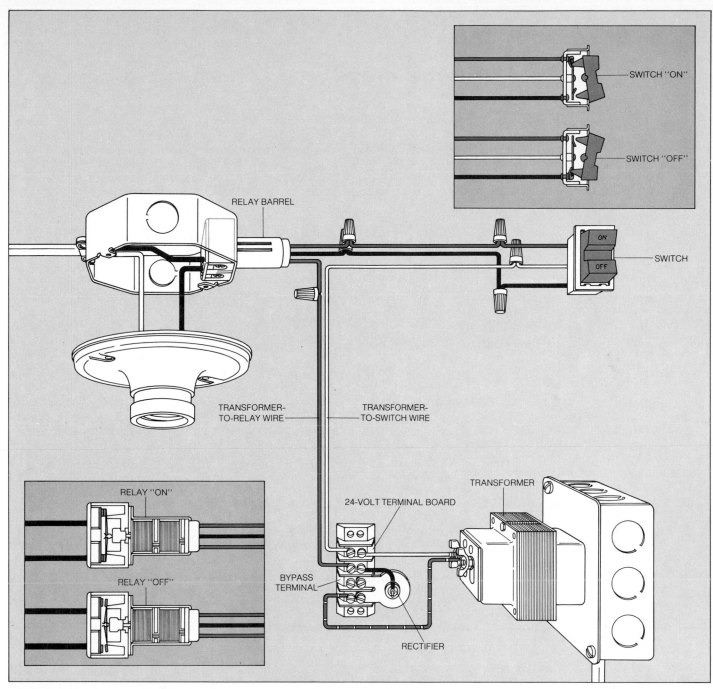

SWITCH "ON"

SWITCH "OFF"

RELAY BARREL

SWITCH

TRANSFORMER-
TO-RELAY WIRE

TRANSFORMER-
TO-SWITCH WIRE

RELAY "ON"

RELAY "OFF"

TRANSFORMER

24-VOLT TERMINAL BOARD

BYPASS
TERMINAL

RECTIFIER

The basic switching circuit. This diagram shows the wiring for a single switch controlling a relay that turns a ceiling light fixture on and off. More complex systems will contain additional elements, but no matter how many switches and relays are added to a system, the wiring always follows the routes shown here.

At the 120-volt Class 2 transformer a 24-volt terminal board connects to switches through a white wire and to relays through a blue one. Closing the "on" contacts of the switch (*inset, top right*) completes a four-part circuit: from the white wire, through a red wire, through an "on" coil

in the relay, and through the blue wire back to the transformer. With current flowing, the "on" coil magnetically draws an iron core inward, snapping a switch-arm to complete the 120-volt circuit of the light (*inset, bottom left*). Closing the "off" contacts of the switch completes a second circuit: white wire, black wire, relay "off" coil and blue wire. The "off" coil pushes the core outward and breaks the light circuit.

At the terminal board, a rectifier protects the relay by modifying the 24-volt alternating current from the transformer. If a relay were fed alternating current continuously (for example, from a

switch jammed in the "on" or "off" position) it would overheat and burn out. The rectifier cuts half the alternations out of the current, reducing heating to an amount that the relay can tolerate. Low-voltage installers distinguish between the rectified and alternating current by wire color: a blue wire recoded with dots of white paint carries alternating current from the transformer to the rectifier; a solid blue wire carries the rectified current on to the relay. The unused terminals on the strip provide a way of bypassing the rectifier; they can be used to power the motor of a master switch (*page 112*), which requires alternating low-voltage current.

The Connections— Transformer, Relay and Switch

1 **Wiring the transformer.** Turn off the power. Wire the 120-volt leads of the transformer into a branch circuit at a 4-inch box and mount the transformer and the box cover plate welded to it on the box. Fasten the terminal-board-rectifier combination nearby, and wire it to the transformer in the pattern shown at right, with a white wire to one of the 24-volt transformer terminals and a blue wire coded white to the other. Connect a length of cable containing a blue and a white wire to the board, still following the pattern at right, and run the cable to the outlet box where you plan to install the relay. Finally, run a length of cable containing white, red and black wires from the relay box to the position you have chosen for the switch.

If your rectifier has a pair of wire leads rather than a terminal board, connect either of the leads to a transformer terminal and the other to the blue wire of the relay cable; connect the white cable wire to the second transformer terminal.

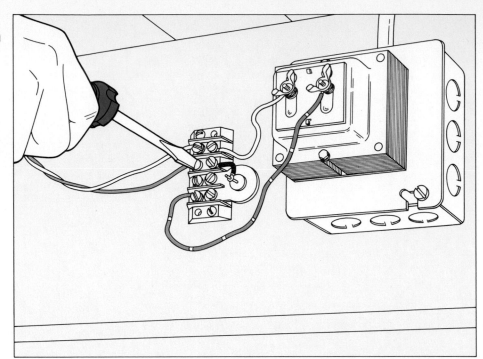

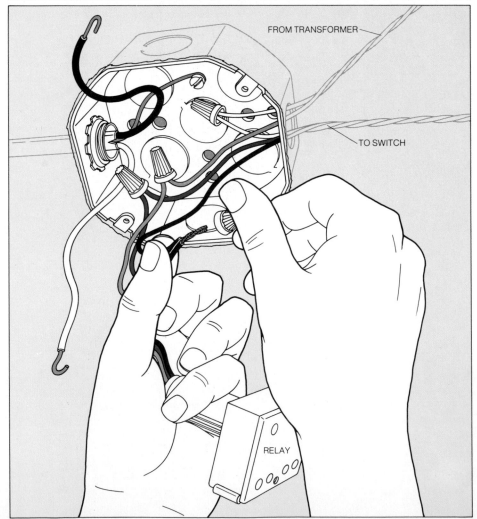

FROM TRANSFORMER

TO SWITCH

RELAY

2 **Low-voltage connections at the relay.** Disconnect the black wires of a light fixture or a receptacle (if you need working space at a fixture box remove the fixture entirely). Pull the two- and three-wire low-voltage cables into the box through a knockout hole. Connect the blue wire from the transformer to the blue lead of the relay and the white wire from the transformer to the white wire of the three-wire cable. Connect the red and black wires of the three-wire cable to the red and black leads of the relay.

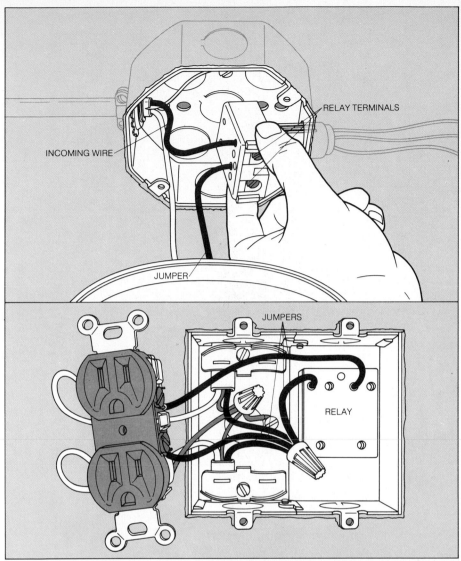

RELAY TERMINALS

INCOMING WIRE

JUMPER

3 **Making the high-voltage connections.** At a box for a light fixture *(top, left)*, connect the black wire of the incoming 120-volt cable and a black jumper or the black wire from the fixture to the high-voltage setscrew terminals at the back of the relay. Push the relay barrel out of the box through a knockout until it snaps into place.

Before mounting a relay in a middle-of-the-run receptacle box *(bottom)*, snap off the small brass tab that joins the receptacle terminals. (If your receptacle does not have a break-tab, replace it with one that does.) Connect the black wires of the incoming and outgoing 120-volt cables to each other and to two short jumpers, one leading to a relay terminal and the other to the bottom terminal of the receptacle; run a third jumper from the other relay terminal to the top terminal of the receptacle. The top receptacle will now be switched by the relay; and the bottom will deliver continuous current.

To wire a relay to an end-of-the-run receptacle, connect the single black wire to one relay terminal and run a jumper from the other relay terminal to a receptacle terminal.

JUMPERS

RELAY

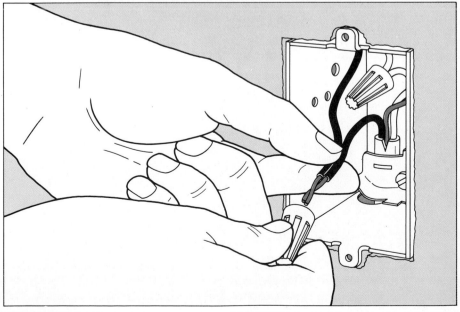

4 **Eliminating an existing switch.** Remove any high-voltage switch controlling the fixture or receptacle, and join the black wires with a wire cap. If you do not intend to use the box for a low-voltage switch cover it with a blank plate.

5 **Connecting the switch leads.** Push the switch through the middle opening of the switch strap until it snaps into place, and connect the three-wire cable to the switch leads—white to white, red to red and black to black.

6 **Mounting the switch.** Install a pair of wall mounting brackets in the hole in the wall (*page 94, bottom*), fold the wires into the wall and fasten the switch in place by screwing the strap to the wall mounting brackets.

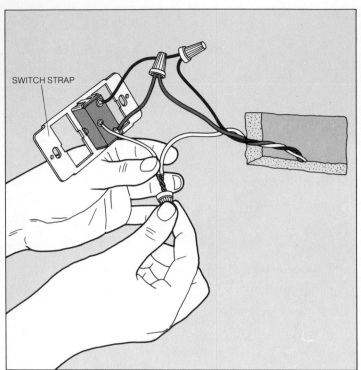

SWITCH STRAP

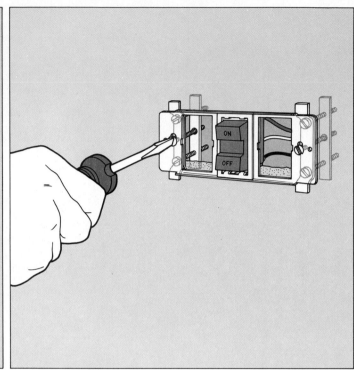

A Telltale Switch

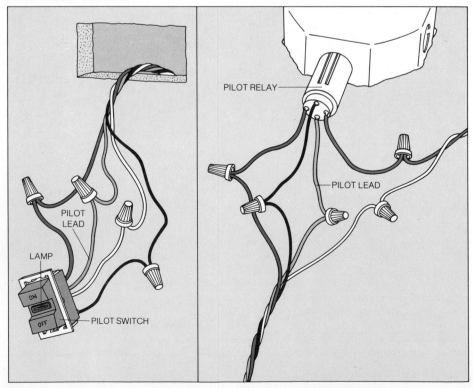

PILOT RELAY

PILOT LEAD

PILOT LEAD

LAMP

PILOT SWITCH

Wiring a pilot light. To monitor a fixture that cannot be seen from the switch location, install a switch equipped with a small lamp that lights when the fixture is on. Run a four-wire cable between the switch and the relay. Connect the white, red and black wires of the cable and switch, and connect the yellow wire from the pilot switch to the fourth wire in the cable (*far left*). At the relay end (*left*), connect the fourth cable wire to the yellow relay lead, and the white, red and black wires as in page 104, Step 2.

Controlling a Fixture from Many Locations

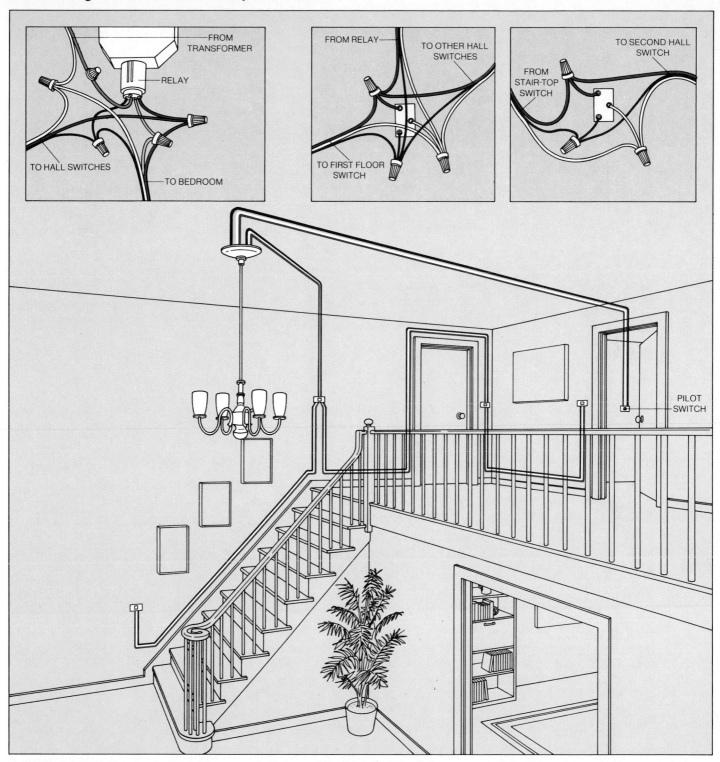

FROM TRANSFORMER

RELAY

TO HALL SWITCHES

TO BEDROOM

FROM RELAY

TO OTHER HALL SWITCHES

TO FIRST FLOOR SWITCH

TO SECOND HALL SWITCH

FROM STAIR-TOP SWITCH

TO SECOND HALL SWITCH

PILOT SWITCH

Multipoint switching. To connect a single relay to a number of switches, run a cable from each switch directly to the relay, or run one cable to the relay and other cables from switch to switch or combine the two methods. A combination of the two methods (*above*) often gives the greatest economy of material and labor. In this example, a hall light is controlled by five switches. One length of three-wire cable runs between the relay and the switch at the top of the stairs. At the switch the three wires of the relay cable are connected to both the switch leads and the wires of two additional three-wire cables—one leading to the switch at the foot of the stairs, the other to the first hall switch. In turn, the first hall switch is connected to the second with another length of three-conductor cable.

The bedroom switch, equipped with a pilot light, is connected to the relay with four-wire cable; the extra wire powers the pilot lamp.

Operating Many Fixtures from a Master Switch

By adding a master switch to the low-voltage remote-control relays described on the preceding pages, you can give yourself space-age command over all electrical devices in your home. When you twist a knob, lights or other devices such as a bathroom heater or a coffeepot will turn on or off anywhere, individually or in any desired sequence. A panel at bedside, in the example illustrated at right, turns lights on or off in all the rooms shown. On a dark winter morning it can turn on, in sequence, bedroom, upper hall, lower hall and kitchen lights, illuminating a path to breakfast. Yet in any room, a light can still be controlled by its own wall switch.

The device that does such master switching consists of two dials that can be rotated to any of 12 positions, which are identified in a dimly illuminated directory of numbered labels above the dials. To turn on any single one of the listed lights (or other devices), you rotate the "on" dial to the light's number and press the dial inward. You can also turn all the lights on or off in quick succession by pressing the dial and then rotating it in one fast sweep. The "off" dial works similarly to put lights out. Pilot lights behind the directory labels tell which lights are on at any particular time.

With more sophisticated equipment such as the motor-driven masters described on pages 112-113, electric timers and photoelectric cells can operate the remote-control switches so that you can not only turn lights on or off from a control panel but also set things up for automatic operation of selected lights, so that they turn themselves on before you return to your house after dark. You can let a timer take responsibility for turning lights off as you retire. In effect, you can have the clock, the sun and a motor master do much of the switching for you.

How a manual master control works. The three-wire cables that radiate from the master control in the bedroom at upper right to lights and receptacles throughout the house are all parts of transformer-switch-relay combinations. When an "on" dial in the control is turned to a specific location—such as the light in the nursery at upper left or the receptacle for the coffee maker in the kitchen downstairs—and pressed, the combination functions as an electrical circuit. To switch on the nursery light, for example, current flows through a white wire from the transformer in the basement to the master control. From there it flows through the red wire of the control cable to the "on" coil of the relay at the nursery fixture, turning it on. To complete the circuit, the current flows from this relay to the transformer through a blue wire. The yellow wire in the control cables is part of another transformer-switch-relay circuit, controlled.

TRANSFORMER

by each relay. When the red wire turns on the nursery light, it also operates an extra switch in the relay, permitting current to flow between the fixture and a pilot light at the control. One control cable, between the control and the master bedroom light, does not contain a yellow wire—the operator at the control panel does not need a pilot light to tell when that relay is on. The black wire in the control cables is an ''off'' wire. Its circuit runs from the transformer (white

wire) to the ''off'' dial of the master control; then through the black cable wire to the ''off'' coil of a relay, which cuts the power to a light or receptacle and also turns off the relay's pilot light; and finally back to the transformer through a blue wire. One blue wire, running between the transformer and the master control, is part of a unique circuit in the system—a circuit that is always in operation. In this circuit, current flows from the transformer to the control through the

white wire and back to the transformer through the blue; at the control panel, it lights a dim locator bulb that stays on all the time.

The circuits for the individual wall switches located near lights and receptacles are the same as the basic switch-relay circuit on page 103. They provide separate control of each outlet from any number of individual switches, in addition to control from the master panel.

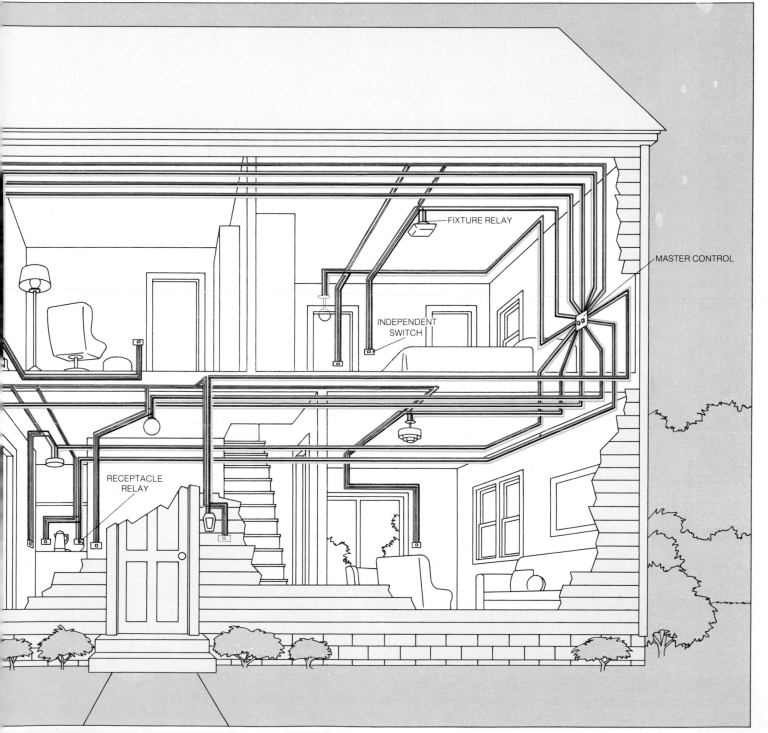

FIXTURE RELAY

MASTER CONTROL

INDEPENDENT SWITCH

RECEPTACLE RELAY

Wiring for a Manual Master

1 **Wiring the switch dials.** After installing the transformer for the switching system and relays for each device to be controlled (*pages 104-105*), gang two standard outlet boxes—use the size that is 3 inches deep—run a two-wire cable from the transformer, run numerically keyed three-wire cables from the relays into the ganged boxes and mount the boxes. Connect the red wire of the relay cable tagged "1" to the red wire at the No. 1 position of the "on" switch; connect the black cable wire to position 1 on the "off" switch. Repeat the procedure with the other relay cables.

Do not cut off any unused wires on the switch dials—you can use them to expand the system—but twist them into neat strands to be tucked in the outlet boxes.

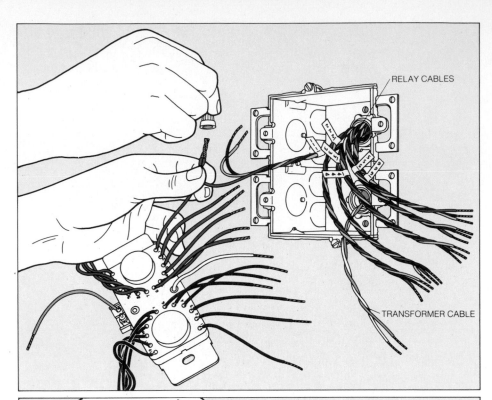

RELAY CABLES

TRANSFORMER CABLE

2 **Connecting the pilot lights.** Matching numbers as you did at the switch dials (*Step 1*), connect the yellow wires of the relay cables to the yellow leads from the pilot-light assembly. Connect the white wire from the transformer to the white leads from the switch and pilot-light assembly, and connect the blue transformer wire to the blue switch-assembly lead.

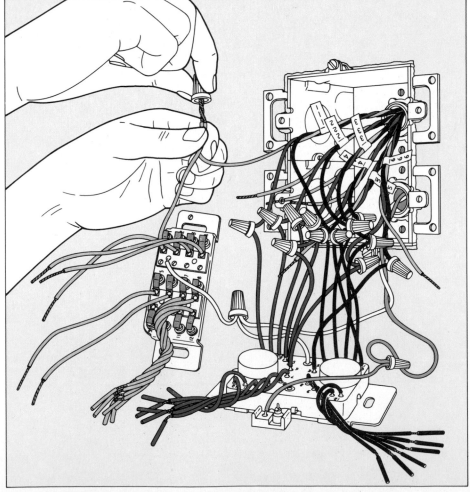

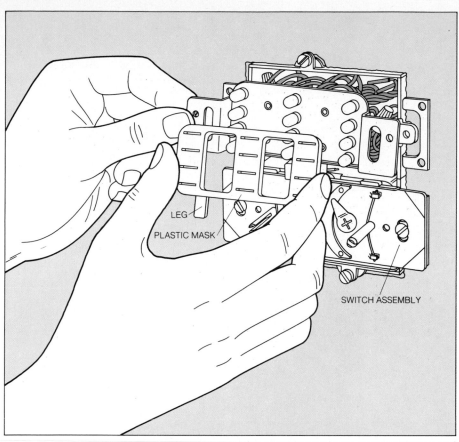

3 **Assembling the switch panel.** Neatly fold the red and black wires into the ganged boxes and screw the switch assembly to the bottom half of the box (the locator light should be at the top of the assembly). Then fold the yellow wires into the boxes and cover the pilot lamps with the plastic mask supplied with the unit; position the pilot assembly by setting the legs of the mask on the top of the switch body and screw the assembly to the top box.

LEG

PLASTIC MASK

SWITCH ASSEMBLY

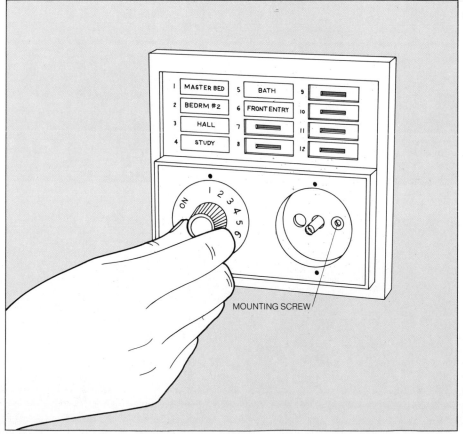

4 **Attaching the cover plate.** Using the adhesive tabs supplied by the manufacturer, label the numbered directory on the cover plate to match your relay numbers. Screw the plate to the switch assembly through holes next to the switch shafts, then turn the left shaft clockwise and the right one counterclockwise as far as they will go. Push the control knobs onto the shafts—''on'' at left, ''off'' at right—lining up the No. 1 knob positions with the indicator dots.

1	MASTER BED	5	BATH	9	
2	BEDRM #2	6	FRONT ENTRY	10	
3	HALL	7		11	
4	STUDY	8		12	

MOUNTING SCREW

Motor Masters— the Ultimate in Remote Control

A motor-master switch uses a small motor to do what a manual-master switch does when pressed and rotated: operate a number of relays one after another. It is a remote-control switch operated by other remote-control switches. Such a system is usually connected to a photoelectric cell or a timer so that a number of selected lights in a house and yard turn on at dusk and off at dawn, or on and off at set hours, while leaving each light subject to individual manual control.

In practice, two motor masters—one for "on" and one for "off"—are installed. The motor is turned on by a 24-volt signal—from any of various switches in any location—to drive an arm that sweeps around 250 contacts; as it touches each it sends a signal of 24 volts to the relay connected to that contact. The arm then turns the motor off and the device awaits the next command.

Motor masters cannot be connected directly to a timer or photoelectric switch because such devices, unlike other remote controls, do not operate with a momentary tripping action. Both close a switch at the time or light condition you choose and keep it closed until the condition set for it to open. Therefore you have to connect them to an electronic device called an interface, which translates the closing and opening of contacts in the control into "on" and "off" signals for the motor masters. In addition a photoelectric cell, which contains a 120-volt internal switch, must be wired through a 120-volt relay to operate the interface. Motor masters and interfaces both use unrectified 24-volt current and must be wired directly to the transformer.

In mass house-light switching, you may want some lights to turn on when the others turn off, and vice versa—for example, you may want yard lights to go on at bedtime when interior lights go off. You can arrange this, in motor masters, by reversing the "on" and "off" leads to the out-of-step lights. Moreover, you can connect individual switches to motor masters so that you can override commands sent by timers or cells.

Wiring for a Motor Master

1 **Connecting the transformer.** After installing the relay and switch circuits for the devices to be controlled, following the procedures on pages 104-106, screw an "on" motor master (red leads) and an "off" motor master (black leads) into a panel box; then connect the blue-and-white leads of each to the blue wire of a cable containing a blue and a white wire. Run the cable to the transformer and connect the white wire to the terminal holding the white wire to the switches. Connect the blue wire to the bypass terminal of the rectifier (*page 103*) or to the other transformer terminal.

Recode the ends of the blue wire by adding white dots. Then run a red-and-black cable from each relay to the motor master, tagging the end with an identifying number.

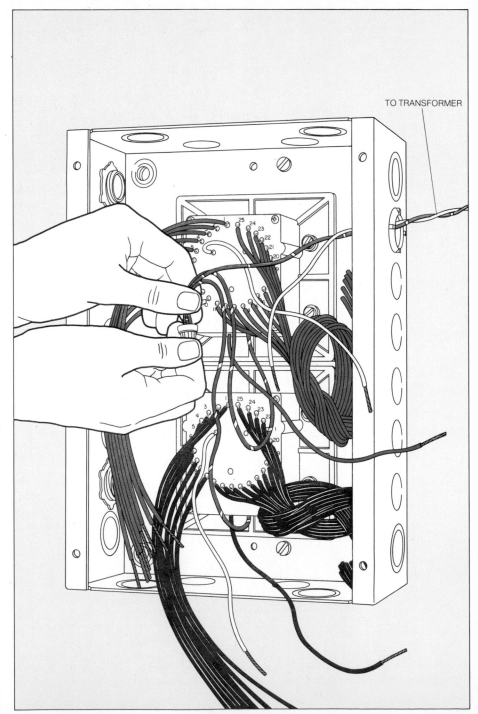

TO TRANSFORMER

2 **Connecting the relays.** Matching colors and numbers, connect the red and black wires of the relay cables to the red and black leads of the motor masters. If you want the motor masters to turn a fixture on while turning others off, connect that fixture's black relay wire to the "on" motor master and its red wire to the "off."

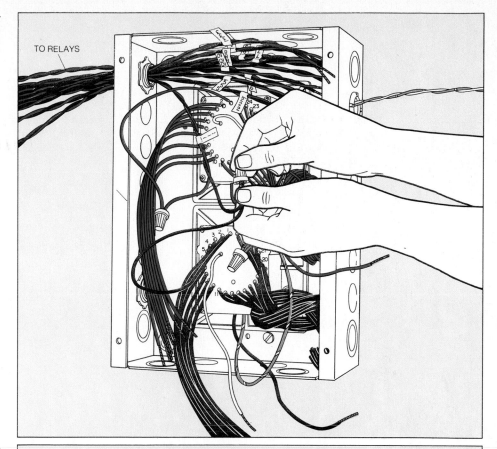

TO RELAYS

3 **Connecting the controls.** Run a red-and-black cable from the interface location to the panel (the interface wiring is shown on page 114). At a convenient place in a living area install a standard 24-volt switch for manual control over the motor masters, and run a white-red-and-black cable from the switch to the panel. (If you want several control switches, wire them as shown on page 107.) Connect the red and black wires of the interface and switch cables to the matching wires of the "on" and "off" motor masters. Connect the switch white wire to the white wires of the motor master and the transformer.

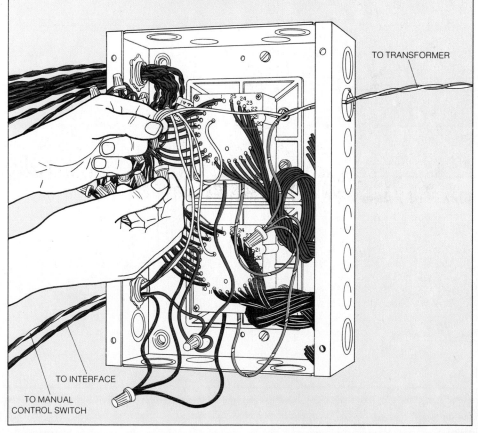

TO TRANSFORMER

TO INTERFACE

TO MANUAL CONTROL SWITCH

Timers and Photoelectric Cells

The interface. Mount the interface near the motor-master panel and make the following connections. For a timer or a photoelectric cell attach two No. 18 wires with 600-volt insulation to the interface screw terminals labeled N.O. CONTROL DEVICE. These wires need not be color-coded—they can be connected in either order. At the terminals labeled MOTOR MASTERS, connect the red wire of the cable from the motor masters to the "on" screw and the black wire to the "off" screw. From the terminals labeled TRANS, connect a white-and-blue cable to the transformer. Recode the ends of the blue wire with white dots. Run the cable to the transformer and connect the white wire to the terminal that holds the white wire to the switches; connect the blue wire to the bypass terminal of the rectifier (*page 103*) or to the other transformer terminal.

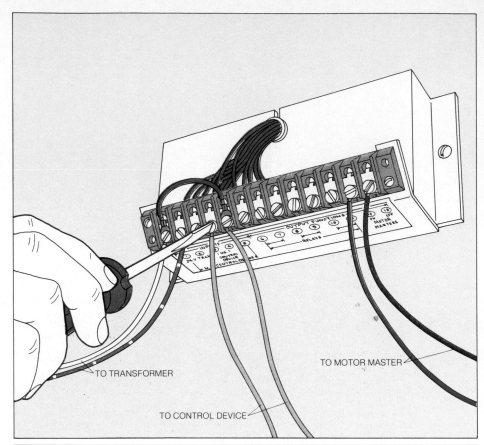

TO MOTOR MASTER

TO TRANSFORMER

TO CONTROL DEVICE

A control timer. Mount a timer near the interface and connect the wires from the interface N.O. CONTROL DEVICE terminals to the timer terminals 1 and 2, then run a 120-volt cable to the timer from a nearby junction box, remove the metal jumper between terminals L and 1 and connect the black cable wire to terminal L, the white wire to terminal X and the ground wire to the green grounding screw.

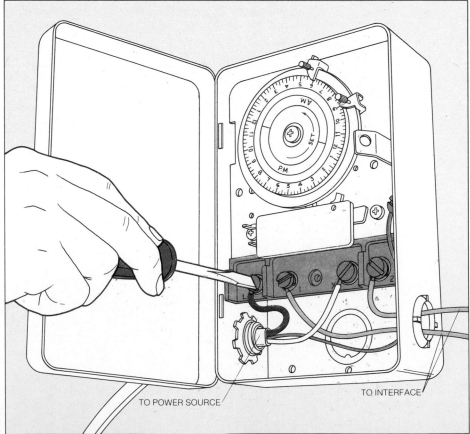

TO POWER SOURCE

TO INTERFACE

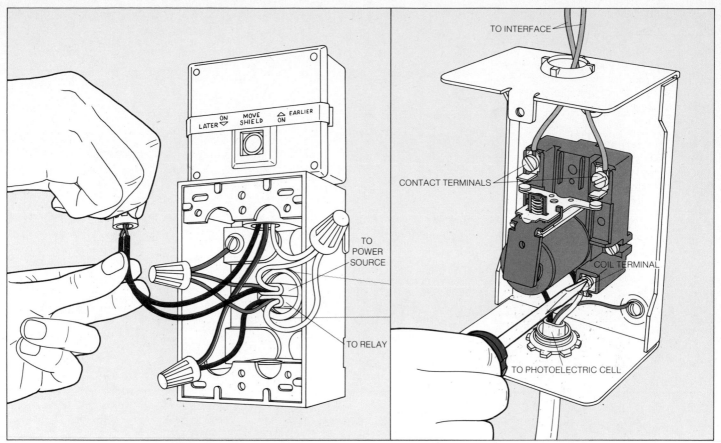

A photoelectric cell and relay. Screw the photoelectric cell to a weatherproof outlet box and mount the box in an open location on an outside wall. Connect the box to the inside of the house with a length of conduit. Feed 120-volt cables from a power source and from the relay location through the conduit and connect the black wire of the photoelectric cell to the black wire of the power cable, the white wire of the cell to the white wires of both cables, and the red wire of the cell to the black wire of the relay

cable; connect the ground wires of the cables to a grounding screw in the box.

In a relay enclosure mounted near the interface, install a general-purpose 120-volt relay that closes contacts when energized. Connect the wires of the cable from the photoelectric cell to the relay-coil terminals in either order, and attach the ground wire to a grounding screw; connect the No. 18 wires from the interface to the contact terminals.

Moving a Garage Door by Motor and Radio

Low voltage can control the high voltages of motors as well as light bulbs; in fact, one of its most effective applications is controlling the 120-volt motor in a garage-door opener. Circuits powered with 24-volt current enable you to open or close the door in about 12 seconds by switches—or, if you prefer, by radio signals transmitted from your car, so that you need not leave the car until you are inside the garage. You can, if you wish, mount a key switch on a post near the driveway, and run its wires in burglarproof conduits. The same circuitry will reverse the motion of a closing door that accidentally strikes a person, pet or object. It can even turn on a garage light and lock the door automatically.

The opener can be installed on almost any garage door, sectional or one piece,

as wide as 16 feet and as high as 8 feet; the only special requirement is a clearance of at least 4 inches between the opened door and the ceiling. The torsion or extension springs that balance the door must be adjusted to let the door move with minimal effort; if the door sticks, the motor will have to be set to supply more power, reducing the sensitivity of the reversing mechanism and making the door potentially dangerous.

Begin by painting the door, if necessary; the weight of a coat of paint affects the spring adjustment. Oil the door hinges and rollers and clean the tracks. Use wires to fasten the lock of the door in the unlocked position. Then adjust the springs so that the door stays open, without support, when 3 feet up.

If your door has a torsion spring—a

windup coil that runs across the top of the doorway—close the door and insert an 18-inch steel winding rod into a hole in the spring winding cone. Holding the rod firmly, loosen the cone's setscrews. Then, using both this rod and a second rod inserted alternately in the cone holes, loosen or tighten the spring. Finally, holding one of the rods, retighten the setscrews. Caution: use rods that fit the cone holes precisely and keep your head and body completely out of the orbit of the rods; a torsion spring is strong and (depending on its design) may unwind in either direction. If you have extension springs—expanding coils *(below)* that run at right angles from the top of the doorway—adjust them in the slack position (that is, with the door open) by taking up or paying out some of the cable.

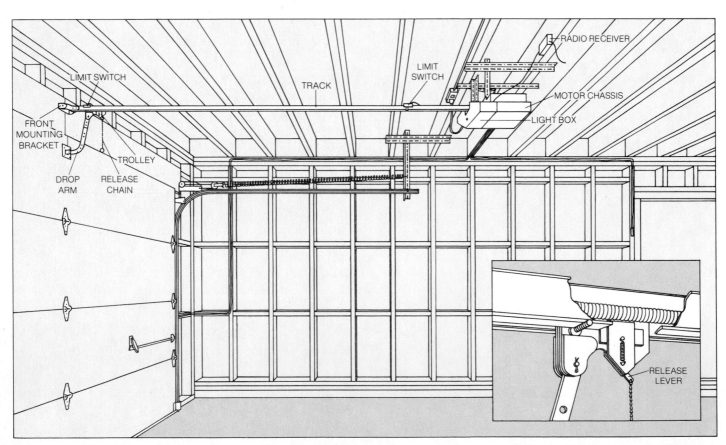

How an opener works. Moving along a metal track, a trolley driven by the motor in the chassis pulls the door open or pushes it shut by a drop arm bolted to the top of the door. Limit switches, struck by the trolley, turn the motor off at the end of a run. In the model shown here, a long screw shaft inside the track *(inset)* moves the trolley; in another type, the trolley is attached to an endless chain, like the chain of a bicycle, looped around a power sprocket at the motor end of the track and a free-turning sprocket at the door end. At the trolley, a pull chain and release lever that free the trolley from the screw or chain make it possible to open the door by hand if the power fails.

Installing the Machinery

1 **Installing the bracket.** Draw a center line on the top edge of the garage door and extend the line vertically up the header over the door, then raise the door until it turns the highest point of its upward arc and set a level from this point to the header. Mark the header at the bottom edge of the level. Lag-screw the front mounting bracket to the header 3 inches above the mark and centered on the line.

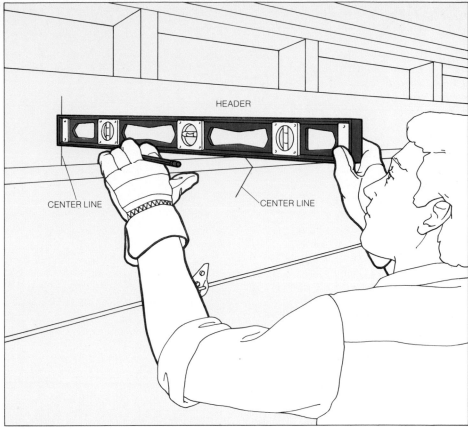

HEADER

CENTER LINE

CENTER LINE

2 **Attaching the track.** Lift the door end of the track up to the front mounting bracket and bolt the track loosely to the bracket. Rest the chassis atop a stepladder and, by inserting wood blocks under it, raise it high enough so that you can open the door without hitting the track.

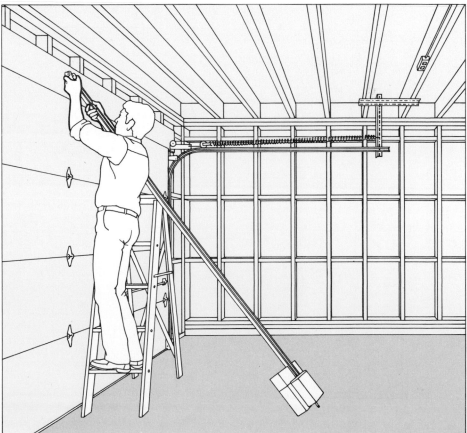

3 **Positioning the chassis.** Working from a ladder, adjust the track sideways to lie precisely above the line on the top of the door, then raise the track high enough to place a brick or a 2-by-4 between it and the door.

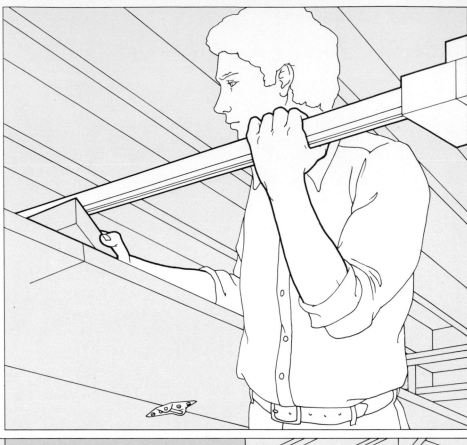

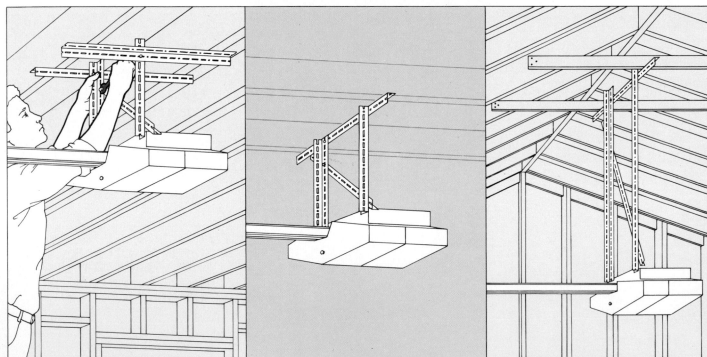

4 **Mounting the chassis.** If your garage has exposed joists, lag-screw a pair of angle irons across the bottom of two joists directly above the boltholes on the chassis (*above, left*), and from these cross supports bolt a second pair of angle irons to the chassis. Bolt a single angle

iron diagonally to the vertical hangers as a brace. If a ceiling covers the joists (*center*), lag-screw a horizontal support to them. If the overhead structure consists of rafters, attach two collar beams to the rafters above the chassis (*right*), lag-screw a horizontal support between them and

use longer vertical hangers, double-braced if necessary to prevent movement in the chassis.

Close the garage door and tighten the nuts on the front mounting bracket. Pull the release chain and slide the trolley close to the door.

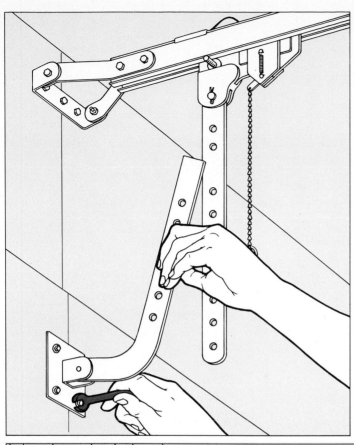

5 **Attaching the drop arm.** Fasten the top half of the drop arm to the trolley with the metal pin and cotter supplied by the manufacturer and let it hang free; then bolt the bottom half to the top of the door at the center line, using carriage bolts in holes that are drilled through the door.

6 **Assembling the drop arm.** Hold the two halves of the drop arm together and, moving the trolley along the track as necessary, find a combination of boltholes that will give the assembled drop arm an angle of about 15° from the vertical. Bolt the parts together.

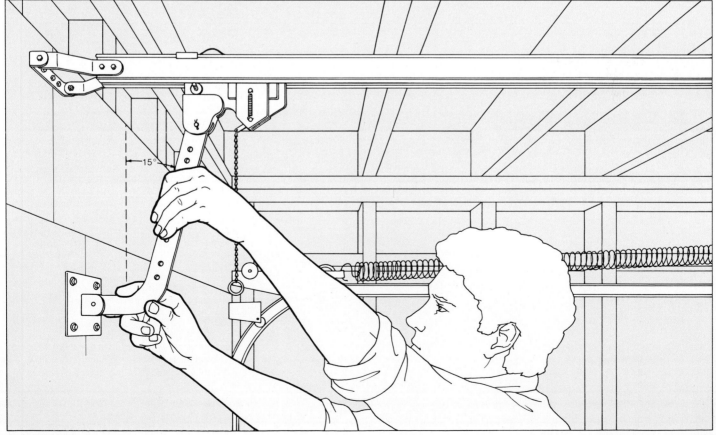

Switches to Control the Door

A garage-door opener draws its power from a standard 120-volt receptacle, which you may have to install yourself at a point convenient to the chassis; do not use an extension cord. The low-voltage circuit, powered by a transformer, includes a relay and switches that operate the opener in a four-stage cycle: stop, up, stop, down. The circuit responds to wall and key switches and a radio receiver located outside the opener mechanism; built into the mechanism are two limit switches and an adjustable reversing device that, in effect, senses when the door has hit an obstacle by noting that the motor is applying more force than is needed to operate the door.

The battery-powered radio transmitter can be moved from car to car, and can be mounted on a dashboard or clipped to a sun visor. It sends a coded signal different from any neighbor's and nearly impossible for a burglar to match; if necessary, you can recode the signal yourself *(opposite, bottom)*. The transmitter will open the door from 75 feet away, but you should not operate the door unless you can see it. A chassis light goes on when you enter the garage in your car or from your house; a time-delay switch turns the light off two minutes later.

1 Installing the switches. Attach a pair of short bell wires to the key-switch terminals, fish the wires into the garage through 1-inch holes drilled in the siding and the inner wall, and insert the switch barrel into the siding hole. Nail the faceplate of the switch to the siding. Inside the garage, connect the wires to a second pair, using a light splice *(inset)* that will pull apart if burglars pry the switch off from outside. Tape the splices. Run the wires to the motor chassis, using insulated staples.

Attach bell wires to the push-button switch, mount it beside a door leading into the house (the switch should be high enough to be out of reach of small children), and run the wires to the chassis. Mount the radio receiver on the garage ceiling, about 3 feet from the chassis.

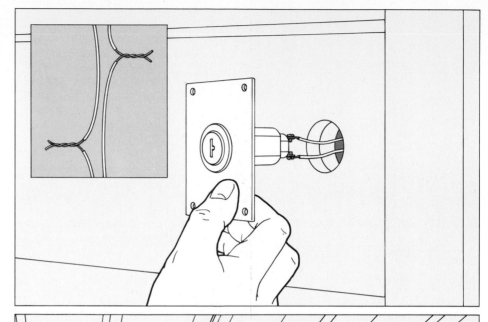

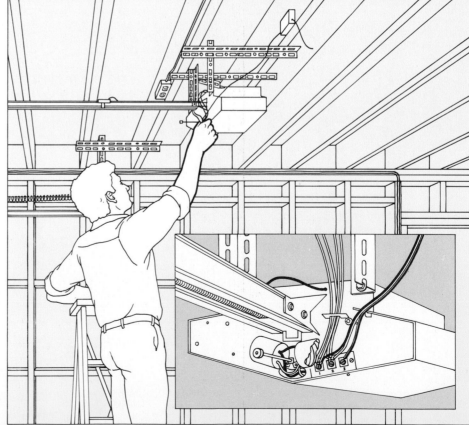

2 Wiring the terminals. Run wires from the three terminals on the radio receiver to the three terminals of the chassis, matching the insulation colors to the terminals' color codes. Connect either wire from the key switch and the push-button switch to chassis terminal 1 and the other two switch wires to terminal 2 *(inset)*.

3 **Setting the limit switches.** Release the motor trolley, close the door and position the movable lever of the "down" limit switch to touch the actuating pin on the trolley; tighten the setscrew at the top of the switch to hold the switch in place. Engage the trolley in its track, open the door and set the "up" limit switch at the other end of the track.

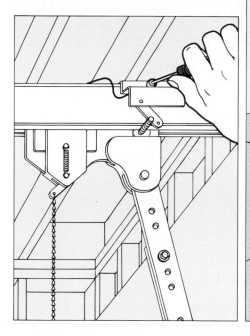

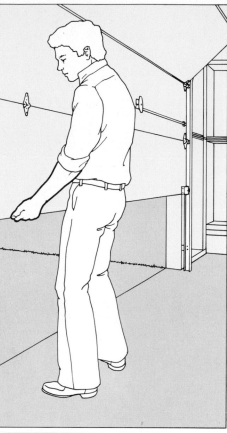

4 **Setting the reversing mechanism.** Have a helper close the door from the wall switch, while you stand at the center of the door opening. When the door reaches a point about 3 feet above the ground, resist its movement with about 35 pounds of lift. (To gauge your lifting force, you may want to hold about 35 pounds of dead weight in your arms before making this setting.) If the door does not stop and reverse, let it slip out of your hands; its downward force is excessive. If it reverses with less than 35 pounds of lift, its force is insufficient. In either case, disconnect the motor at the chassis and reset the adjustment screw on the reversing mechanism. Repeat the test until the setting is correct.

Professionals use 35 pounds of pressure, but you may employ less force by starting with insufficient downward force and gradually tighten the adjustment screw until a force barely necessary to close the door is reached.

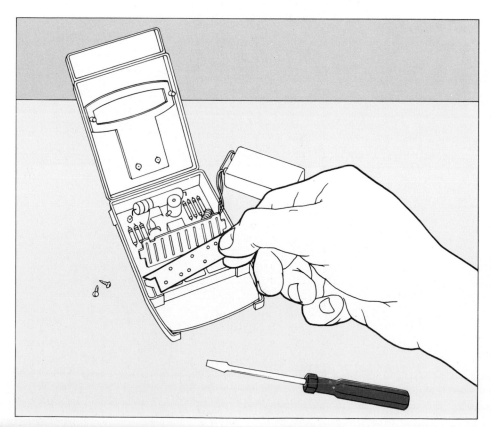

Changing the Radio Code

The turn of a card. If your radio receiver operates randomly from stray signals—or if you find that it has the same signal code as a nearby door opener—change the code preset by the manufacturer. To recode the widely used punch-card system shown at left, open the transmitter box, remove the screws that hold the card in place, turn the card over from top to bottom and replace the screws. Follow exactly the same procedure in the receiver. If trouble persists, turn the cards over sideways and, if necessary, once again from top to bottom.

Other transmitter and receiver models are coded and recoded by a row of nine small slide switches. Each switch can be moved to either of two positions with the tip of a pencil to create hundreds of code sequences.

Outdoor Lighting: Tiny Twinkles Around the Yard

Low-voltage outdoor lighting, with thin cables and inexpensive fixtures, makes it easy to have exactly the yard lighting you want. The fixtures connect to the cables with outdoor pin connectors or weatherproof splice caps. They can be clipped to tree branches or staked into the ground; waterproof models can safely be set underwater, in a pool or fountain. And, because the connecting cables need not be deeply buried (they are generally hidden along hedges, fences and walks, stapled to tree trunks or tucked into a shallow slit cut with a lawn edger) the lights can be pulled up and moved at will.

Most outdoor lights come in a kit containing a 12-volt transformer sealed in a weatherproof box, a length of cable and six to 12 fixtures, although you may prefer to assemble the elements separately with a variety of fixtures.

The transformer can be mounted on a wall near an outdoor receptacle. If this location requires more than 100 feet of cable to supply the fixtures, you must allow for voltage drop—a slight consumption of current by wires that will dim distant lights. Heavier cable solves the problem but it is harder to hide and less flexible than the standard sizes. Many installers prefer to mount the transformer on a post well away from the house and run a 120-volt line to it. In a very large yard, it may be best to mount two transformers at widely spaced locations for separate groups of fixtures.

Installing the transformer. Mount the wall bracket supplied with the transformer at least 12 inches above the ground on a wall or post, and within 6 feet of a receptacle; then push the transformer onto the bracket.

Place the fixtures and string the low-voltage cable from the transformer, leaving enough slack to follow walkways and flower beds. Leave 12 inches of additional slack at each fixture.

Clipping a fixture to a cable. If your fixtures have pin connectors, make a 3-inch cut in the cable sheathing to separate the conductors, but do not strip any insulation from the conductors; lay the wires into the slots at the base of the fixture, press them firmly onto the contact pins and screw the protective cap into place.

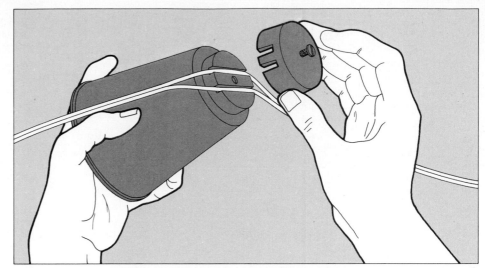

Using splice caps. For a fixture with wire leads, cut the cable at the locations you have chosen, strip ⅜ inch of insulation from each cable and fixture wire, and twist each fixture wire to conductors from each end of the cut cable; crimp a splice cap over each three-wire connection with a multipurpose tool *(right)*.

Partially fill the insulator with silicone caulking compound and slip the insulator over the splice cap *(far right)*. Caution: be sure that the cap is embedded in the compound.

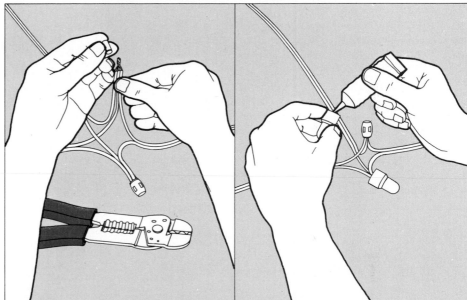

Hiding the cable underground. Push an edging tool into the ground with one foot, rocking it back and forth to exert a downward and sideward pressure. Overlapping strokes will cut a narrow slit; tuck the cable into the slit and step along the top to press the turf back together.

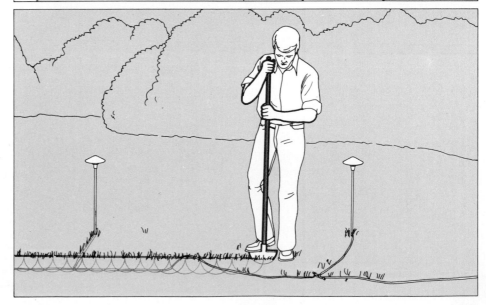

Calculating Loads According to the Code

To decide whether a new circuit will overload a service panel, most electricians use a few rules of thumb and a great deal of guesswork. When in doubt, they put in a new, larger panel, perhaps unnecessarily. You can, if you like, use the same approach—some of the rules are given on page 10—but you can obtain a more precise result and avoid needless labor by applying rules and formulas in the National Electrical Code.

When checking the effect of an added circuit, you can take advantage of the fact that the electrical loads in your house do not draw power all at once or at full capacity. Some pairs of appliances—air conditioners and heaters, for example—are never used simultaneously. The Code defines such loads as "noncoincident" and allows you to include the larger of them in your calculations and to ignore the smaller one.

Other loads, such as electric ranges and the components of a central-heating system, seldom run at full capacity: the circuit breaker and cable that serve these appliances must be large enough to handle the entire load, but the Code lets you

discount or "derate" part of the loads when you calculate the total capacity required of the service panel.

Finally, since only a fraction of the lights and appliances in your home draw power at any one time, the Code specifies 10,000 watts (10 kilowatts) as a basic power requirement and lets you discount 60 per cent of the remainder of the load when calculating the total requirement.

When deciding on what new circuits to install, allow for certain Code requirements for various types of circuits:

☐ General lighting circuits. While the Code requires three watts of power per square foot of floor space, it does not limit the number of fixtures and receptacles these circuits serve. In practice a 20-ampere circuit should not supply more than 16 outlets or a 15-ampere circuit more than 12. The lighting circuits can supply stationary appliances—refrigerators, small air conditioners and the like—but the sum of the appliance ratings must not exceed half the circuit rating, and no single appliance can exceed 80 per cent of the circuit rating. (Note: In Canada lighting circuits are limited to 15 amperes

and a total of 12 fixtures and receptacles.)

☐ Small-appliance circuits. The Code requires at least two small-appliance circuits, and most homes have more than two; a circuit generally supplies two or three receptacles. These circuits are designed to serve high-wattage kitchen appliances in a dining room, family room, pantry or kitchen; they cannot be used to serve receptacles in other rooms or lighting fixtures in any room. The restrictions on large appliances described above apply to small-appliance circuits.

☐ Special-purpose circuits. The Code requires only one special-purpose circuit—a 20-ampere, 120-volt laundry circuit for a washing machine—but most homes have several others, for clothes driers, refrigerators and freezers, ovens, ranges, air conditioners and other large appliances. Such circuits can be powered by either 120 volts or 240 volts, and can be wired either directly to an appliance or to a receptacle within 6 feet of the appliance. Generally, each circuit supplies only one or two appliances. The wire sizes for a variety of appliance amperages are given in the chart at the bottom of this page.

A Wire for Every Circuit

Amperage	Wire size (AWG)
15	No. 14
20	No. 12
30	No. 10
40	No. 8
55	No. 6
70	No. 4

Choosing the right wire. The National Electrical Code requires that you match the rating of a circuit to the size of the wire it serves: a wire that is too small for its load can heat dangerously without tripping the breaker.

The Right Cable for Every Service

Service-panel amperage	Copper cable	Aluminum cable
100	No. 4	No. 2
125	No. 2	No. 1/0
150	No. 1	No. 2/0
200	No. 2/0	No. 4/0

Sizing service entrance cable. This chart correlates the rated amperage of a service panel with the size of copper or aluminum cable required to serve it. The choice between copper and aluminum is a matter of convenience and cost: copper is expensive but easy to bend and connect; aluminum, the more common material, is much cheaper but far less flexible. In service entrance cables both materials are

considered equally safe, but aluminum must be coated with anticorrosive paste before making connections and requires lugs and connectors marked AL or AL/CU.

In Canada copper is preferred, and the requirements differ slightly: No. 2 AWG for 100 amperes, No. 1 for 125 amperes, No. 1/0 for 150 amperes and No. 3/0 for 200 amperes.

Branch-circuit Amperages

Type of circuit	Amperage
General lighting	15 or 20
Small appliance	20
Clothes washer	20
Non-motor appliance	Name-plate rating
Motor appliance	125% of name-plate rating

The right amperage for each circuit. The National Electrical Code sets minimum capacities for certain circuits, depending on their purpose: general lighting, small appliances or clothes washer. The Code does not specify the amperage for circuits serving large appliances—space heaters, electric ranges and the like—but the capacity of their circuits must be greater than the amperages indicated on the appliance name plates. For motor-driven appliances such as an air conditioner or a stationary power saw, the circuit amperage is calculated by multiplying the name-plate amperage by 125 per cent to allow for the current surge when the motor starts. If a large motor-driven appliance—one drawing three amperes or more—is on a circuit with other devices, this adjustment only affects the largest motor; use the name-plate rating of the others.

Calculating the Load on a System

	Formula	Example	Result
Heating and air conditioning load	65% of central electric heat	65% × 14,000	9,100 watts
	100% of air conditioning	100% of 9,700	9,700 watts
Other loads	General lighting—3 watts per square foot	3,000 sq. ft. × 3 watts	9,000 watts
	Small appliance circuits—1,500 watts each	3 circuits × 1,500	4,500 watts
	Laundry circuit—1,500 watts or name plate	1,200-watt washer	1,500 watts
	Clothes drier—5,000 watts or name plate	5,500-watt drier	5,500 watts
	Other major appliances	1,200-watt dishwasher	1,200 watts
		4,500-watt water heater	4,500 watts
		7,400-watt range	7,400 watts
		9,500-watt oven	9,500 watts
	Total of other loads		**43,100 watts**
Derating the total load	First 10 kW of other load at 100%	100% of 10,000	10,000 watts
	Remainder of other load at 40%	40% × 33,100	13,240 watts
	Heating or air conditioning load, whichever is larger		9,700 watts
	Total derated load in watts		**32,940 watts**
Load in amperes (watts ÷ volts)		32,940 ÷ 240	**137 amperes**

Calculating the amperage for a house. This chart is actually a work sheet used to calculate the size of the service for a typical home. The first column lists the categories in which calculations are made; the second lists the formulas for making the calculations; the third gives examples of specific figures for appliances and circuits; and the fourth shows the results of applying the formulas to these figures. If your home has a three-wire service of 100 amperes or more, you can use the same method, substituting your own figures in columns 3 and 4.

For heating and air conditioning, the formulas are simple: 65 per cent of the central electric heating load, in watts or kilowatts (kw), and 100 per cent of the air-conditioning load. If you have fewer than four separately controlled space heaters, treat the sum of their wattages as central heating; if you have four or more, add all of them to the "other loads" category rather than to the heating load.

When calculating square footage for the general-lighting circuit requirements, count all living spaces, including attics and basements; exclude crawl spaces, open porches and garages. For the laundry circuit, use 1,500 watts or the name-plate rating of your washing machine, whichever is higher; if you have a clothes-drier circuit, rate it at 5,000 watts or 125 per cent of the name-plate rating. Under "other major appliances" include the wattages of every major appliance you own and of any additional appliances or circuits you plan to install.

To determine the total electrical demand, add together 100 per cent of the first 10,000 watts of "other loads," 40 per cent of the remainder of "other loads," and 65 per cent of the heating load or 100 per cent of the air conditioning, whichever is larger. Divide this figure by 240 volts to convert it into amperes. In this example the total is 137 amperes, and a 150-ampere service would be large enough.

Picture Credits

The sources for the illustrations in this book are shown below. The drawings were created by Roger C. Essley, Fred Holz, Georgine E. MacGarvey, Joan S. McGurren and Jeff Swarts. Credits for the pictures from left to right are separated by semicolons, from top to bottom by dashes.

Cover—Fred Maroon. 6, 8, 9—Fil Hunter. 11 through 15—Peter McGinn. 16 through 19—Whitman Studio, Inc. 20 through 23—Gerry Gallagher. 25 through 31—Whitman Studio, Inc. 32 through 35—John Massey. 36, 37—Walter Hilmers Jr. 38—Fil Hunter. 41 through 47—Fred Bigio from B-C Graphics. 48, 49—Walter Hilmers Jr. 50 through 53—Nick Fasciano. 54 through 57—John Massey. 58 through 63—John Sagan. 64—Fred Maroon. 66 through 71—Walter Hilmers Jr. 72, 73—Eduino J. Pereira. 74 through 81—Whitman Studio, Inc. 82 through 85—Great, Inc. 86, 87—John Massey. 88 through 95—Fred Bigio from B-C Graphics. 96 through 99—Walter Hilmers Jr. 100—Fred Maroon. 103 through 115—Vantage Art, Inc. 116 through 121—Peter McGinn. 122, 123—John Massey.

Acknowledgments

The index/glossary for this book was prepared by Mel Ingber. The editors also wish to thank the following: David M. Baum, Dynalectron Corporation, Alexandria, Va.; Robert Beall, Walter C. Davis & Son, Alexandria, Va.; David Beatty, David Beatty Stereo Hi-Fi, Kansas City, Mo.; James Bennett Jr., B & C Construction Company, Oxon Hill, Md.; George Birch and Peter Ristau, Buhl Electric Company, Vienna, Va.; W. Randolph Black, Gene Divers and Larry Miller, Virginia Electric and Power Company, Alexandria, Va.; H. M. Bowling, Tennessee State Department of Insurance, Nashville, Tenn.; George Brady, Audio Associates, Falls Church, Va.; David Brock and Bill Cowherd, Marvin C. Cowherd Company, Falls Church, Va.; Bruce Bunker and Peter F. Monahon, Meyer-Emco, Washington, D.C.; Charles W. Carr, Miniature Lighting Products, Port Richey, Fla.; A. Catalano and Reuben Holmes, Joseph M. Catalano Company, Falls Church, Va.; Steve Coen, Underwriters' Laboratories, Chicago, Ill.; Richard F. Conn, George Washington University, Washington, D.C.; Buddy Davis, Raymond Davis and Randy Million, Davis Antenna Company, Waldorf, Md.; Edison Electric Institute, New York, N.Y.; Marian Finney, Kenneth G. Newton and Rickey E. Nowlin, NuTone Division, Scovill Company, Cincinnati, Ohio; D. C. Fleckenstein, General Electric Company, Fairfield, Conn.; Martin E. Harp, Electrical Inspection Department, City of Alexandria, Va.; Ken Hightower and Don Scheetz, Square D Company, Park Ridge, Ill.; William B. Howell and David Marshall, Capital Lighting and Supply, Alexandria, Va.; Alan Isicson, Certified Electronics, Alexandria, Va.; M. J. Kornblit and Earl W. Roberts, General Electric Company, Plainville, Conn.; Walter Krimont, Store Radio, Inc., Washington, D.C.; Walter Kromer, San Jose, Calif.; Harry Kuntzelman, Underwriters' Laboratories, Melville, N.Y.; James Lemnah and Charles E. Reutter, McGee and Company, Washington, D.C.; William J. Locklin, Loran, Inc., Redlands, Calif.; J. C. Lyttle, Ontario Hydro, Toronto, Ontario; Tom Mather, James G. Biddle Company, Plymouth Meeting, Pa.; Roland Morgan, Best Antenna Services, Arlington, Va.; Reginald Pollack, Great Falls, Va.; Thomas J. Pugh and William T. Shepard, Wiremold Company, West Hartford, Conn.; Hans Rabong, Winegard Company, Burlington, Iowa; Alan Reed, Daniel Woodhead Company, Northbrook, Ill.; David R. Rick, Nokeville, Va.; Yank Rowan, Graybar Electric Company, Lanham, Md.; L. H. Sessler, Bell Telephone Laboratories, Whippany, N.J.; Sal Soscia, General Electric Company, Providence, R.I.; Barry Staley, Door Systems, Inc., Lorton, Va.; John Sternberg, Radio Shack, Alexandria, Va.; Robert Stewart, Rush Electric, Alexandria, Va.; George Vallance, E. M. Logan & Sons, Alexandria, Va.

The following persons also assisted in the preparation of this book: Barry Fishler, David Garfinkel and David Schrieberg.

Index/Glossary